THE
CONSERVATIVE
ENVIRONMENTALIST

THE CONSERVATIVE ENVIRONMENTALIST

COMMON SENSE
SOLUTIONS FOR A
SUSTAINABLE FUTURE

BENJI BACKER

SENTINEL

SENTINEL
An imprint of Penguin Random House LLC
penguinrandomhouse.com

Sentinel and colophon are registered trademarks of Penguin Random House LLC.

Most Sentinel books are available at a discount when purchased in quantity for sales
promotions or corporate use. Special editions, which include personalized covers, excerpts,
and corporate imprints, can be created when purchased in large quantities. For more
information, please call (212) 572-2232 or e-mail specialmarkets@penguinrandomhouse
.com. Your local bookstore can also assist with discounted bulk purchases using the
Penguin Random House corporate Business-to-Business program. For assistance in
locating a participating retailer, e-mail B2B@penguinrandomhouse.com.

Graphs by RPrime unless noted otherwise.

Library of Congress Cataloging-in-Publication Data
Names: Backer, Benjamin, author.
Title: The conservative environmentalist: common sense
solutions for a sustainable future / Benji Backer.
Description: New York: Sentinel, 2024. | Includes bibliographical references.
Identifiers: LCCN 2023042705 (print) | LCCN 2023042706 (ebook) |
ISBN 9780593714003 (hardcover) | ISBN 9780593714010 (ebook)
Subjects: LCSH: Backer, Benjamin. | Environmentalists—
UnitedStates—Biography. | Climate justice.
Classification: LCC GE56.B33 A3 2024 (print) |
LCC GE56.B33 (ebook) | DDC 333.73092 [B]—dc23/eng/20231117
LC record available at https://lccn.loc.gov/2023042705
LC ebook record available at https://lccn.loc.gov/2023042706

Printed in the United States of America
1st Printing

BOOK DESIGN BY TANYA MAIBORODA

CONTENTS

Evolution, Not Revolution

Above New York City's Union Square looms an eighty-foot-wide digital display known as the Climate Clock that is counting down, second by second, how much time the planet has before it hits 1.5 degrees Celsius of global warming above pre-industrial levels. This number marks Earth's so-called *deadline*, the designated threshold at which catastrophic changes in the environment will become irreversible. At this point, sea levels are predicted to engulf all coastal cities, ice sheets will be completely lost, and all animals will begin disappearing.

Sounds ominous, right? It's supposed to. Alarmism sells, and a realistic discussion about climate change doesn't, so I've learned.

You need only to look at the content on your social media feed to know that fear is the number-one tool used by political influencers and news reporters to elicit thousands of vicious tweets, disparaging comments, and shock-faced emojis. With outspoken voices on the political right calling green energy "deadlier than the coronavirus" and those on the political left predicting a fast-approaching end-of-the-world date, it's understandable that fear and outrage about our future characterize our generation.

I don't agree with their tactics, *or* the idea that Earth will be an apocalyptic wasteland by 2050, but alarmists are right that we have a problem on our hands. While I've clashed with my fellow conservatives over whether climate change is real, the science—as well as the empirical evidence of record-breaking temperatures and the increased frequency of natural disasters—tells us definitively that the Earth really is warming up. And though the planet has been steadily increasing in temperature since the Industrial Revolution, we've seen *significant* increases in recent history.

Over the past seven decades, carbon dioxide levels have climbed nearly five times as fast as in the decades prior. To put those changes in some historical context, the amount of rise in carbon dioxide levels since the late 1950s would naturally have taken somewhere in the range of 5,000 to 20,000 years. We've managed to do it in about sixty years. At the same time, the rate of warming averaged 0.14 degrees Celsius per decade. The nine years from 2013 to 2021 rank as the warmest ever recorded. The rapid rate of temperature rise over such a short period of time points to only one thing: greenhouse gases.

That said, reacting with despair is the last thing we should be doing. In times of trouble, the best thing we can do is think logically and act methodically. Instead, the conversation has been dominated by Greta Thunberg, Alexandria Ocasio-Cortez, and the Green New Deal, and calls for panic-and-despair-driven measures like the immediate and total ban of fossil fuels. Unlike Greta and many well-known climate activists, who call for governments around the world to take extreme and drastic measures that will impact our liberty and economic security, I'm a conservative. I want to change the narrative surrounding climate change—and spread the good news that there is so much we can do, and are already doing, to fix the climate without sacrificing the things that are important to us: economic security, an open and free market, and our liberty. I want to take care of the environment with local solutions, capitalist incentives, and innovation— all of which are compatible with conservatism. "Conservation" and "conservative" are cognates, after all.

Eyebrows are always raised when I tell people I'm a conservative environmentalist, but the reality is that there are many of us. In 2019, 82 percent of 18- to 35-year-olds polled indicated that climate change is important to them, including 77 percent of right-leaning and 90 percent of independent respondents. Many rural, conservative Americans whose hometowns are among regenerative farms, sustainable agriculture, and land conservancies see caring for the environment as a natural way of life. Further, some of the most climate-friendly states, including Maryland, Indiana, New Hampshire, and Georgia, which all boast some of the highest rates of emissions reductions in the country, are led by Republican governors. Yet when it comes to finding their voice in a political climate driven by fear and catastrophic thinking, most conservatives don't know where to start.

It wasn't always this way. The truth is, conservatives have a track record of enacting pro-climate policies, even if you'll never hear about them in the mainstream media. Theodore Roosevelt earned his name as the Conservation President when, during his time in office, he created 150 national forests, 55 federal bird reserves, and 5 national parks, altogether protecting approximately 230 million acres of public land. In 1970 President Nixon created the Environmental Protection Agency through executive order, which has since served as a cornerstone in the protection of human health and the environment. In his State of the Union address that year, he declared, "The great question of the Seventies is . . . shall we make our peace with nature and begin to make reparations for the damage we have done to our air, to our land, and to our water?"

Two years later, frustrated with the lack of movement on the eighteen environmental bills he hoped to pass, he delivered an urgent message to Congress: "These problems will not stand still for politics or for partisanship." When all was said and done, Nixon ended up leaving a legacy of legislation like the National Environmental Policy Act, Clean Air Act, and Endangered Species Act.

During his term, President Ronald Reagan set aside over 12.5 million acres of public land and helped mitigate the national ozone crisis by establishing the international Montreal Protocol, which succeeded in phasing out harmful chemicals from common household products. This policy was lauded as one of the greatest environmental achievements of the twentieth century.

George H. W. Bush, who succeeded Reagan, pledged that he

would be remembered as the Environmental President. Most famously, he advocated for and signed the 1990 Clean Air Act Amendments, the most expansive environmental regulatory legislation in the nation's history. The act served to eliminate the dense smog present in nearly every major US city. It also tackled the problem of acid rain caused by sulfur dioxide emissions from coal-burning power plants, which had been devastating the lakes and forests of the northeastern part of our country. Even more notable than the success of Bush's environmental efforts was his economically creative, albeit at times controversial, approach. Instead of the traditional command-and-control strategies that had dominated environmental policymaking, he used a cap-and-trade mechanism that harnessed the power of the market. The government wouldn't tell polluters how to lower emissions, but simply imposed a cap on them each year.

Eight years later, George W. Bush would carry on his father's environmental legacy, signing historic bipartisan legislation that accelerated the cleanup of toxic brownfields (land formerly used for development that has significant environmental contamination) as well as implementing the Healthy Forests Initiative, which helped restore valuable woodlands and rangelands and reduce forest fires. The younger Bush also designated nearly 140,000 square miles of the Northwestern Hawaiian Islands as a Marine National Monument.

Conservative climate-action champions continued to make headlines on the state level too. Republican California governor Arnold Schwarzenegger signed the California Global Warming Solutions Act, making California the first state in the nation to cap greenhouse gas emissions. The governor is famous for declaring, "California will not wait for our federal government to take strong action on global warming. . . . International partnerships are needed in the fight against global warming and California has a responsibility and a profound role to play to protect not only our environment, but to be a world leader on this issue as well." He would make good on his word by leaving an impressive legacy of environmental firsts, including the creation of the Western Climate Initiative, the world's first International Carbon Action Partnership (ICAP) in collaboration with European Union countries, other US states, and Canadian provinces, and the first-in-the-nation mandatory Green Building Standards Code (CALGreen).

My goal is to carry on this long tradition of conservative environmentalism. This book is the culmination of seven years of on-the-ground conservative climate advocacy, focused on shifting the narrative from partisanship, divisiveness, and rhetoric into one of solutions, actions, and results. I'm writing it because I'm tired of the tribal approach to climate politics that pits well-meaning people against one another in an ongoing battle of moral and intellectual self-righteousness. I'm writing it because I'm optimistic about our ability to rationally address climate change.

GROWING UP IN the Midwest, the great outdoors was an integral part of my childhood, and I struggled to formulate my own beliefs about climate change while campaigning for conservative candidates like Governor Scott Walker and presidential candidate Mitt Romney.

When I moved to Seattle for college, I searched Google for a right-leaning environmental advocacy group to join—but found nothing. Not only that, but I also had no idea what types of solutions a conservative could support.

Then one afternoon, I was sitting in my business sustainability class when I recalled some advice I'd held on to from a few mentors: "If you're going to start something, start it when you're young." Before my professor even finished her lecture, I bought the domain name for the "American Conservation Coalition" and sent out a tweet asking people to join me—though I wasn't quite sure what they'd be joining me in yet. I just knew doing nothing—and knowing nothing about an alternative approach—wasn't an option.

For the next four years, in addition to my studies, I volunteered sixty to eighty hours a week, pushing the American Conservation Coalition (ACC) off the ground, while still in school. Some days the work was endless, and most of the time I had no idea what I was doing, but for the first time, an alternative climate movement was being born. That hard work paid off, and now the ACC boasts 25,000 members in every corner of the United States.

At its inception eight years ago, ACC's membership was nearly 100 percent conservative. Today, our membership is made up of roughly 60 percent conservatives and 40 percent independents and liberals. This degree of bipartisanship among young people proves what I've been saying all along: climate change is a human issue, not a political one. And its solutions lie in the common-sense approach of improving the environment and reducing carbon output, an approach that transcends political ties. To many who have used climate change to advance their party's agenda, this idea is a tough pill to swallow.

Perhaps the greatest paradox of all is ACC's message of hope amidst the cacophony of skepticism and despair. To many in the climate community, it's a given that the Earth is doomed for imminent destruction—and the only question that remains is which part of the planet will be hit first. I'm not negating the seriousness of the issue. However, I believe that by approaching short-term issues with practical solutions, sensible policies, actionable steps at an individual level, and a long-term commitment to innovation, we can bend the curve of destruction and significantly improve our prospects.

Over the years since I founded the ACC, I've traveled all over the country—to coal mines, oil fields, wetlands, timberlands, and solar and wind farms. I've seen firsthand how American conservation and innovation are making a difference in countless communities, with local public-private partnerships leading the way. I've witnessed our growing new movement: conservatives showing how deeply they care about the environment. Most importantly, I've seen how communities innovate climate solutions that work for them, regardless of any political narrative. Whether it's duck hunters protecting precious wetlands, urbanites instituting important recycling programs, or Midwestern engineers working tirelessly to improve wind turbines, the small, local wins add up to a big difference.

I've made it my mission to alleviate people's fears about climate catastrophe so that conversations about climate can become less focused on reactive rhetoric and more on productive solutions. I meet with environmental entrepreneurs, clean energy engineers, and fossil fuel leaders involved in powerful solutions that most news stories and social media feeds don't even mention. I listen to the underrepresented voices of coal miners, field workers, and farmers who have been dismissed by the urban elite as uncaring or unevolved. And I meet with policymakers concerned about the environment who also care about preserving our liberty, our ability to earn a living, and our national security.

The efforts of the ACC are already paying off. Our members have helped to pass important legislation like the Great American Outdoors Act of 2020, which funded the debt-ridden National Park System, and the Energy Act of 2020, which funded domestic clean energy businesses. This legislation was the first bipartisan environmental action Congress had seen in years. We've also hosted dozens of summits and rallies for the climate with Congress members from both sides of the aisle. Our recent efforts to bridge the gap led to the passing of the Growing Climate Solutions Act, a sustainable agriculture bill, which passed with massive bipartisan support in the Senate.

And we're just getting started. In the following pages, you will discover dozens of unexpectedly effective environmental realities that go against common thinking. *Did you know smartly producing more domestic fossil fuels would actually help with global warming in the short term? That reducing food waste is the fastest way to lower carbon emissions right now? How better forest management could reduce carbon emissions more than switching to electric vehicles? Why the current plan to convert solely to solar and wind power is doomed to fail? What trophy hunting has to do with helping the environment?* By answering questions like these, I'll dispel myths perpetuated by sensationalist reporting and political agendas while also exposing immediate and long-term solutions that work for our people *and* our planet.

This book serves as a model for the future of environmentalism—and is largely centered around the pillars of "The Climate Commitment," an alternative climate framework recently launched by ACC. You'll find the themes of the framework's pillars throughout all chapters of the book—the chapters get their names from the core action words in The Climate Commitment.

The Climate Commitment

To summarize, the plan includes the following six pillars of action:

1. **Streamline:** We can only tackle climate change as quickly as we can build modern, clean energy infrastructure. By getting unnecessary government regulations out of the way, we will allow America's entrepreneurs to build cleaner and faster, while maintaining high environmental standards and protecting our beautiful landscapes.

2. **Unlock:** Rather than relying on competitors for the building blocks of the clean energy economy, we should unlock America's own natural resources—while maintaining the highest environmental standards in the world.

3. **Protect:** By taking a community-first approach, we protect the important—and vastly different—cultures and economies across our nation. Additionally, as we build a cleaner future, we must ensure the process is also prosperous, to pro-

tect the communities and livelihoods affected by our decisions.

4. **Compete:** By using the power of the marketplace and capitalism, we can use our nation's competitive spirit to lead the
world in emission-reducing technologies.

5. **Conserve:** One of the most affordable and immediate steps
we can take to be more efficient is restoring our country's rich,
beautiful ecosystems. Most of the funding and infrastructure already exists for better forests, wetlands, grasslands, and
coastal management.

6. **Innovate:** By boosting research and development and investing in next-generation agriculture, conservation, nuclear, hydrogen energy, battery storage, carbon capture technology,
and more, we will continue to make long-term progress.

The Climate Commitment—and this book—is the product of
the ACC's work to cut through the noise, engage conservative
voices on solutions authentic to their values, and build a coalition
of people on both sides of the aisle who are willing to fight—
together—for a better future.

In Chapter One, we'll dive into how this issue became so di-
vided, the communities we're missing from the dialogue, and the
problems with the current environmental narrative. In Chapters
Two through Seven, we'll discuss the surprisingly simple solutions
in front of us—most of which have been ignored by the mainstream environmental movement. These solutions are the ones
we *must* pursue in the short, medium, and long term if we hope to
tackle environmental challenges in a sensible way. We can't afford

to ignore the climate *or* our own interests as a nation. Fortunately, there's a way to protect both.

I hope, after reading this book, you'll be inspired to join me and the thousands of others who care about the environment by becoming a member of the ACC. Protecting the environment is not something we can do without the other half of the nation. It's time to put petty grievances aside and work together to do something great—after all, that's how this country was founded.

1

Overcoming
Our Political Divide

magine, for a moment, that you are a farmer. If it's winter, your day begins before sunrise when you check the temperature in your greenhouse to ensure your indoor crops aren't overheating. If it's summer, the rest of your morning may consist of cutting, raking, and drying the hay you'll use to feed your livestock. If it's planting season, you may spend your entire day seeding, fertilizing, and spraying your fields; and if it's harvest season, you'll be spending long, long hours in a combine reaping your corn, soybeans, cotton, and other invaluable resources Americans need for everyday living. Each of these complex processes requires its own piece of expensive, specialized equipment, and all of it runs on—you guessed it—fossil fuels.

Not only does all farming equipment run on fossil fuels, but the pesticides and fertilizers that have increased crop yields tenfold and allowed us to feed the country at record-low costs are made with them. The small planes used to efficiently spray large fields run on fossil fuels, the trucks farmers use to haul equipment run on fossil fuels, and the farmer's house runs on fossil fuels. Your fuel costs have always been high, but you've been able to make your business work thanks to the subsidies that the government provides to the farmers who feed the nation.

Now take a moment to think about how you, a farmer, might feel about the government ripping away the subsidies you've relied on for years to run your business, and introducing sweeping, top-down regulations to combat carbon emissions. Your fuel costs double or triple because of a new tax that penalizes users of fossil fuels. You're also paying a higher electricity bill as electricity companies raise rates to cover the installation of thousands of new EV charging stations across your state. If you're a corn farmer, you'll soon find yourself with a lot of excess crop on your hands, as 40 percent of our nation's corn goes to making the now-useless ethanol that powers gas engines. With soaring costs and a superfluous crop, you decide to sell the farm that's been in your family for generations to a foreign solar company who will build thousands of solar panels on what used to be crop fields. As more American farms go under, more and more food is imported from foreign countries, instead of being planted, grown, and harvested on American soil. On top of all this, you find yourself being ridiculed as backwards and ignorant for daring to voice any kind of opposition against the new climate change legislation, even though you support green policies and alternative energy. Truthfully, farmers have more cause than most to be concerned about climate change—as our weather becomes more volatile, the business of planting and harvesting becomes less and less predictable.

As you can see, top-down energy solutions simply do not work in a country as big and diverse as America. If you believe in our democracy—and I very much do—then you believe that we need to implement solutions that work for *everyone*, not just the folks for whom buying a new EV and adding solar panels to their house is convenient. Introducing sweeping, one-size-fits-all policies that harm local communities and ruin the livelihoods of thousands of rural dwellers is simply anti-democratic and anti-American. In thinking about how to solve climate change, it's imperative that we give farmers, oil field workers, and coal miners a voice too.

If you stop to think for one moment about the plight of a fossil fuel–dependent worker in our current political climate, you begin to see that the political division surrounding this issue is material, and not just rhetorical. Fossil fuel–dependent workers are concerned about their very survival in a world that is rapidly transitioning to alternative energy sources—they aren't backwards, ignorant, or stupid, as much as Alexandria Ocasio-Cortez might say they are. Politicians and the media have effectively made green energy into something you're for if you're a Democrat and against if you're a Republican. This simplistic distinction only serves to divide the American people and further the political careers of a few bad actors on both sides of the aisle.

In a cheap effort to gain voters, conservatives and liberals have both adopted equally dramatic approaches toward climate change. Oklahoma senator James Inhofe's 2015 snowball-tossing stunt on the Senate floor—dismissing global warming—still makes the rounds on the internet. We've heard outspoken individuals such as Tucker Carlson and President Trump repeatedly purport that climate change is *one big hoax*. Just as ridiculous was New York senator Charles Schumer blaming climate change for tropical storms occurring in the southeast during that region's normal hurricane season, and climate activists blaming Texas's 2021 snowstorm on it. (Just to be sure we're on the same sheet of paper, a cold, dry snap is exactly the type of weather that global warming is expected to create *fewer* instances of.)

When I share my beliefs about how climate change is affecting the environment, I brace myself as I anticipate a blow from the Left or the Right. I'm not the only one. Politicians and business leaders dread the shame-and-blame game too, and adopt positions they don't really believe just to avoid attacks. One politician explained to me that even engaging in rational discussion about climate change would appear to their constituents as the equivalent of joining the other side.

By the time President Trump took office, partisanship regarding environmental stewardship had been well established. A far cry from the work of his Republican predecessors, the focus for President Trump was to hammer environmentalists and roll back environmental policies. At the same time, climate change was soon lumped in with other important issues that dominated the airwaves such as the #MeToo movement, pandemic protocols, and the Black Lives Matter movement, issues that further polarized the nation. If you glance at just about any election map since 2016, you'll notice that there's a stark red/blue divide—the urban areas are blue, and the rural areas are red. People assume that anyone in a blue area cares about climate change, and anyone in a red area doesn't. While it's undeniably true that rural areas tend to skew conservative, it's absolutely *not* true that these voters don't care about the environment.

The Urbanite Bias

Growing up in a small Midwestern city nestled among dairy farms and forests as far as the eye can see, I witnessed rural Americans hard at work. People there wake up at the crack of dawn, and business owners and laborers get their hands dirty to feed the rest of the country. Our local cheese stores shipped to all parts of the States and were the source of many people's livelihoods. I assumed that everyone else recognized rural America's value too.

When I moved to Seattle to go to college, I learned how wrong I was.

Three-quarters of Washington lies east of the Cascade Mountains, and is too frequently dismissed and mocked by those living along the Pacific coast. When I first heard my friends from Seattle, Tacoma, and Bellevue belittle their eastern neighbors, my stomach turned. (To this day, it still does.) Even though residents of Eastern Washington grow the food we eat and provide the energy powering our homes, many of them lack the bachelor's degree that makes them relevant in the eyes of their urban neighbors.

I often think about the highly skilled agrarian and mechanical workforce in my home state of Wisconsin, who needed very specific training to do their jobs effectively and safely. While a liberal arts education is not high on the priority list of many rural workers, the technology needed to remain on the cutting edge is. Unfortunately, even basic modern conveniences that the rest of the country takes for granted are still lacking in many parts of America. About 28 percent of rural Wisconsinites, for example, still don't receive high-speed internet service, making it impossible for rural residents to work from home or start a new business. This absence of a service that urbanites take for granted leaves many rural families one giant step away from the economic opportunity that is so readily available everywhere else. There are simply more resources available in urban areas to transition to a clean energy future—even for low-income residents. As I mentioned, a push for an electric vehicle transition is easier for a city dweller. (Think of a low-income resident of Chicago using public transportation to get around town versus a low-income farmer who needs to own a truck to haul materials that dictate their economic livelihood.)

Being lower-tech often means that rural America cannot financially keep up with the speed at which the rest of the country is accelerating. For example, California's aggressive plan to ban the sale of all gas-powered vehicles by 2035 spells hardship for lower-income rural areas. Online behemoths such as Amazon, Target, and Walmart will be able to pivot with every change imposed by plans like these, but Dollar Trees and other local discount stores will suffer devastating effects as delivery prices soar, as will the communities who depend on them. The closing of one of these smaller stores may seem like no big deal to someone living in a coastal city. But to its employees, it may mean months, even years of searching for new employment that will pay their family's bills.

Meanwhile, power plant workers, miners, and farmers see the fruits of their sweat and soil placed at the feet of the urban elite who consume their products and services and offer insults in return. Many millennials and Gen Zers can appreciate a good meal from a farm-to-table restaurant or a made-in-the-USA pair of jeans and are even willing to pay a pretty price for these items. However, few give much thought to the incentives and policies that will keep these local businesses afloat. The mantra *Know where your food comes from* is important for reasons beyond a person's individual well-being and should apply to products from industries beyond local agriculture. Yet how many of us give these farmers and manufacturers a second thought once our purchase has been made?

The *adapt or die* mentality of some climate alarmists poses a very real threat to the day-to-day survival of many hardworking Americans. What the urban and suburban elite consider improvements in quality of life are often actually giant setbacks for the rest of the country.

It's no surprise when rural communities are resentful toward their urban counterparts for ignoring their needs when it comes time to vote. Even more disturbing is the neglect they feel when it comes to their states' elected representatives. I've talked with countless rural residents who feel forgotten by politicians who live and work in metro areas hundreds of miles away. In rural places and small towns, people feel they are not getting their fair share of decision-making power. Nor do they feel their concerns, which greatly diverge from those of urban residents, are even being heard.

Although opinions of former president Donald Trump vary widely, there's no denying he managed to capture the hearts of rural Americans by visiting their communities and addressing their frustrations and concerns head on. Even Trump supporters who were well aware of his shortcomings or admitted to disliking his leadership style believed he was willing to stand up to the urban elites and fight for their needs. His appeal to rural Americans was not about facts or particular policies per se, but about the overall message that he understood them and (essentially) had their backs.

In contrast, most other candidates have repeatedly failed to offer America's heartland a clear vision that speaks to their concerns. Promises on the campaign trail that fail to come to fruition have left rural voters disappointed and bitter. To add insult to injury, for years, rural people have heard from the media that they are voting "against their own self-interest" when they elect Republicans, or that they vote the *wrong way* because they are uneducated. These arrogant messages are not easily forgotten—nor should they be.

Trump's promise to bolster goods-producing industries and provide more jobs hit an overwhelmingly strong chord with the working class, especially those living in rural America, where economic damage has hit the hardest in recent years. Compare this message to President Biden's recent address to a group of New Hampshire miners who faced job insecurity and economic setbacks. The president's advice to them? Start learning code: "Anybody who can go down three thousand feet in a mine can sure as hell learn to program as well . . . Anybody who can throw coal into a furnace can learn how to program, for God's sake!" Understandably, the president's reductive comments about an entire industry of workers were met with silence from his insulted audience.

Thus, the false narrative of *two Americas* presents two societies so fundamentally separate and different that they hardly belong to the same world. It frames urban Americans as forward-thinking and productive, and rural Americans as closed-minded and stagnant. The reality is if our leaders are not listening to farmers, foresters, miners, or other rural-based workers and equipping them with the resources they need, the resulting economic harm done to them will have a negative impact on urbanites as well.

Further, it's simply untrue that rural dwellers don't care about the fate of our planet. Putting the term *climate change* aside, rural communities, whether skeptical or not, have always known how to care for their surroundings better than the people who live far away from nature. Their personal stake in protecting their backyards informs the way they behave, and they have learned from experience what works and what doesn't work to keep nature thriving. Some of the cleanest areas in the country belong to private landowners who don't rely on somebody else to make decisions about their property. For farmers, the stakes are high. If they can't keep the land healthy, they will lose their livelihood and not be able to feed their families.

Consider the following scenarios: A city dweller who takes up very little land, uses mass transportation on his daily commute to work, and drinks from reusable water bottles believes he is doing the most to fight climate change. Meanwhile, a suburbanite who drives an EV, composts and recycles religiously, and chooses paper instead of plastic bags at the grocery store believes her contribution counts most. Lastly, a farmer who cares daily for his surrounding land, which not only produces the crops that feed the rest of the nation but also naturally sequesters carbon, might feel that he in fact is the best steward of the planet.

Which one of these three people is doing the best job taking care of the planet? The answer, of course, is all of them. Each person, whether urban or suburban or rural, is partially correct in their approach to protecting the environment. Likewise, Americans living in different parts of the country have a lot to teach one another about effective climate change solutions.

While we cannot deny real political, geographic, and economic trends, we also cannot let them obscure the fact that urban and rural communities need one another to thrive. Nowhere is this interdependence clearer than in our fight against climate change. And nowhere is our failure to recognize that interdependence clearer than in the Green New Deal.

Why the Green New Deal Just Won't Work

The Green New Deal (GND) is the perfect example of how far this issue has pushed the American population to both extremes. Introduced to Congress by Representative Alexandria Ocasio-Cortez in 2019 and championed by Vermont Senator Bernie Sanders in his 2020 presidential campaign, it is the ultimate package of top-down regulations, grand gestures, and sweeping reforms that would result in devastating effects on rural, poor, and carbon-intensive communities. The Green New Deal's giant scope includes many solutions that pertain to climate change—and many that don't—that are as demanding as they are vague and as unrealistic as they are unaffordable. Ultimately, the Green New Deal is a prime example of the urban-rural divide: a bill made by urbanites that fails to take into account the expertise and concerns of rural communities across the country.

To illustrate, let's take a look at a summary of the proposal's main talking points:

- Achieve net-zero greenhouse gas emissions by 2030
- Create millions of union jobs in green technology
- Invest in infrastructure for electric vehicles and for solar and wind power
- Protect quality of life for under-represented communities, such as Indigenous peoples and communities of color

While these may sound like goals environmentalists could share across political boundaries, a deeper dive showcases why it's been so polarizing. In essence, it calls for drastic change and an inflexible methodology with very little nuance. The Green New Deal favors only a handful of solutions that work for only a small percentage of wealthy Americans, especially mandating electric vehicles (EVs), solar, and wind-power as the sole focus areas—which aren't even guaranteed to reduce *global* emissions (more on this later). No one likes drastic change, especially when it threatens their livelihood—or limits access to a resource essential to daily living, such as energy. Fear has been the driving force not only for the outspoken Left, who have taken over the conversation with calls for revolutionary reform, but also for those on the right who are leery of such a radical approach and simply run the other way. The Green New Deal's greatest promise—reaching zero emissions by 2030—is, oddly enough, the least supported by the science. A simple analysis using Project Drawdown's emissions categories shows the proposals within the Green New Deal only cover 31 percent of the emissions categories related to climate change. Now, let's do some math. The United States contributes 13.5 percent of global emissions. Assuming the Green New Deal is *100 percent effective* with what it claims it can do, the policy itself would only reduce emissions by 4.34 percent worldwide (31 percent × 14 percent). This is all for the cost of $93 trillion— a price tag larger than the size of the entire global economy. Meanwhile, the plan almost completely ignores countless market-driven and nature-based solutions that would not only boost our economy but also create easy carbon-reducing wins.

The chart below breaks down the GND's proposed budget. As you can see, 17 percent of the total is allocated to EV subsidies alone, which will only benefit a very small portion of the population. At the same time, only 3.6 percent is going to green farming practices, 1.42 percent to wetlands and coastal restoration, and a measly .01 percent to forest management. These nature-based solutions that enhance the Earth's natural carbon sequestration capability, and are an undeniable, proven strategy in climate action, are entirely overlooked by the Green New Deal. As mentioned, even if completely successful, the Green New Deal would only reduce emissions by about 4.34 percent. For an expected

GND INVESTMENT ALLOCATION

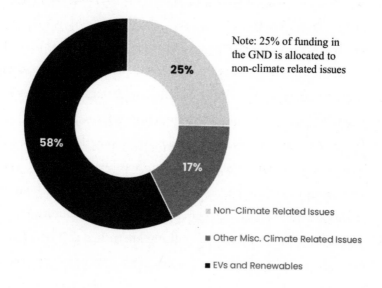

25%

58%

17%

Note: 25% of funding in
the GND is allocated to
non-climate related issues

▪ Non-Climate Related Issues

▪ Other Misc. Climate Related Issues

▪ EVs and Renewables

price tag of $93 trillion, does that seem to be the best use of our dollar?

To put this into even better perspective, it costs between $5 and $100 per ton removed of carbon when reducing the size of forest fires. When you calculate what the Green New Deal would cost per ton of carbon removed, it's a colossal number, between $10,000 and $15,000. And that's assuming *all* global emissions were eliminated under the plan, which, as we've covered, isn't anywhere near possible.

It's no surprise that the Green New Deal's unrealistic call to reduce greenhouse gas emissions to net zero by 2030 has raised eyebrows across the political spectrum. Even House Speaker Nancy Pelosi, a Democrat, has critiqued the GND for such overly ambitious goals, referring to it as the "green dream or whatever they call it." And former energy secretary Ernest Moniz (serving under President Barack Obama) remarked to NPR, "I'm afraid I just cannot see how we could possibly go to zero carbon in a ten-year time frame." He is not alone, as more and more people from both sides begin to question the GND's viability and its unfoundedness in scientific logic or economically sound principles.

In addition, the GND would massively strengthen the federal government's control over activities such as how people produce and consume energy, harvest crops, raise livestock, build homes, drive cars, and manufacture goods. The rural communities that will be most affected by GND reforms are treated as if they hold no stake in the policymaking conversation, so of course, they view the plan as a threat. When a farmer in the Midwest who grows and hauls produce all day long (which you and I consume, by the way) hears about a plan to do away with the fossil fuels that heat their home and power their equipment, he runs the other way. Wouldn't you?

What the GND fails to account for is that even a small win is still a win. In fact, as we have witnessed historically time and time again, lasting change begins with small increments. For example, if a community that has previously relied on coal eventually switches to natural gas, which is 50 percent cleaner, that's a win. On an even smaller scale, if a recent college graduate trades in her 1990 SUV for a newer car that gives her double the gas mileage, that's a win too. Yet, let's face it, these smaller efforts do not make for big headlines.

Climate solutions themselves need to work for rural Americans and their unique set of needs. In 2021, I attended the UN Climate Change Conference (COP26) in Glasgow, Scotland, where I had the chance to speak with legislators, elected officials, and environmentalists from all over the world. I spoke with one Malaysian leader about the rural and urban divide in the US, hoping to gain insight into whether this was a global issue and how he was handling it in his country.

The conversation became eye-opening for me as he explained that in his country in general, gaining support on environmental issues was a challenge because constituents had so many unmet basic needs to address first. Having come from a home where food, shelter, clean water, healthcare, and transportation were a given, I hadn't realized what a privilege it is to even get to care about the environment.

By and large, most American urban areas—even the poorer neighborhoods—have at their disposal the same daily provisions I had growing up. But this is not the case in many rural communities, where employment is scarce, energy is less reliable, and infrastructure is less developed compared to their urban counterparts. Simply tagging on a handful of socio-economic policies to a set of costly and ineffective environmental solutions to win the votes of underserved constituents is not the answer. With a whopping 40 percent of Americans, many of whom live in rural areas, unable to come up with $400 in an emergency, we can't expect to put people's livelihoods at odds with environmental progress.

We need to work harder to understand the concerns of our rural communities and advocate for local approaches that take into account the values of both rural and urban regions. Federal mandates and top-down regulations don't work. Rural voters simply want to elect leaders who see a future for their communities in which their energy is cheap and clean, their people have good jobs, and their local economies are strong.

Charting a New Path Forward

Many mainstream youth climate activists demand sweeping changes that involve banning coal, fossil fuels, and combustion-engine cars, and believe that EVs and wind and solar power are the ultimate trifecta of saving the planet. This oversimplified solution-set disregards the future of low-income, carbon-reliant households, 54 percent of which rely on natural gas, propane, or fuel oil to heat their homes. They also undervalue the opportunities for economic growth in local climate-friendly innovation and land conservation. Most importantly, their fearmongering tactics have only served to incite anger and anxiety in their followers and further deepen the partisan divide.

Meanwhile, climate change deniers disregard scientific discovery and call climate action a leftist conspiratorial agenda. Even more extreme are fossil fuel maximalists such as Alex Epstein who, believing that we can have our cake and eat it too, insist that the convenience, affordability, and reliability of oil and gas far outweigh their carbon and methane emissions impact. They acknowledge that climate change is real, yet suggest that if we do nothing to reduce our consumption of fossil fuels, we can still become lower-carbon by actually using fossil fuels as part of the solution. But the idea that what got us into this mess can also get us out of it is short-sighted and illogical. It excludes the possibility that new, cleaner technologies could become just as affordable and abundant as oil and gas, once they receive the same amount of subsidies that support their improvement.

Instead, politicians and activists cling to a handful of ideas that more and more evidence shows us just won't work. We've all heard the expression that insanity is to keep performing the same action over and over again, expecting different results. Conversely, keeping an agile mindset when it comes to policymaking is a healthier, more resilient approach. A confident doctor welcomes a second opinion and is humble enough to change their course of action because they want what is best for the patient.

As we've covered, a top-down, one-size-fits-all, urban-only approach to climate change will not work. I believe local communities are resilient, yet climate change affects each American neighborhood, city, and state differently. Therefore, the policies we put into place impact these communities differently too, depending on many geographic and economic variables. Americans deserve solutions that work for our own localities and climate action that brings affordability regardless of their income levels or where they live.

You may have noticed a few themes developing when it comes to the solutions we should be pursuing. Namely, those solutions have to be balanced, targeted, measurable, adaptable, and supported by real people. To be more specific:

a. A solution is **measurable** when we can use numbers to quantify its success. Much of the time, the government throws a lot of money at a problem with no real way to measure whether the solution worked. This is foolish for obvious reasons—if we implement a policy that seeks to remove carbon from the air, we should be able to determine how much carbon has been removed. (The Green New Deal, by the way, has no measurable metrics.)

b. **Balanced** solutions have cross-partisan support. As I've mentioned, we can't solve the climate challenges we face with just one party. If key stakeholders aren't at the table, the policy risks missing key elements. That's why we need conservatives *and* liberals, urban *and* rural dwellers, activists *and* corporations to weigh in on these issues. Together, we can create real change.

c. **Targeted** solutions have clearly defined objectives, deliverables, and goals. An example illustrates why this is important: The Clean Water Act of 1972 was just thirty-three pages long, passed quickly with bipartisan support, and resulted in a dramatic decrease of pollution in America's waterways. In contrast, the Inflation Reduction Act of 2022 came in at 787 pages, and while it contained some good climate policies, its unrelated name and confused goals have made it a prime target for reversal by its opponents.

d. **Adaptable** solutions can move with the times, no matter how much they're a-changin'. New technology, new science, new circumstances, new people in power—adaptable policies remain useful no matter what's happening in the world.

e. Solutions that **are supported by real people** are popular with general voters *and* the communities that are most impacted by them. Whether it was the Civil Rights Movement or ACC's push for National Park funding, long-lasting solutions are rarely at odds with the people themselves.

Using these criteria for policy, we can create the kind of progress that benefits the environment *and* all of us.

It's time to stop kicking the carbon can down the road and begin to take swift action—action that *also* coincides with conservative values. To do this, our generation needs to continue to bridge the partisan divide by putting aside theatrics, having sensible conversations, and demanding balanced policies. We must create a robust movement in the "perilous middle" to find results. We also need to embrace job opportunities based in climate solutions. We can do all these things—as long as we're not prevented from enacting practical solutions by bureaucratic red tape.

But first, we need to address another split that threatens to impede true climate progress: the role of government. Let's take a closer look.

The 411 from Chapter 1

- **Environmental issues weren't always so polarizing. However, between the media's incentive to polarize us, conservatives embracing climate denial, and liberals proposing radical policies, there are many reasons why this changed.**

- Throughout history, Republican presidents championed the biggest environmental policies of all time—in partnership with Democrats.

- When Washington, DC, makes climate policies, they often use a top-down approach biased toward resource-rich urbanites, leaving out the interests of other critical communities—especially the rural ones we rely on each day.

- The Green New Deal is a major roadblock to taking climate action. And worse, even if it worked with 100 percent efficacy, it would reduce emissions by only 4.34 percent for the cost of $93 trillion.

2

Streamlining the Complicated Role of Government in the New Green Economy

A s a kid, I hated eating vegetables. Whatever kind was on my plate, I refused to take a bite. Potatoes were too bland, the smell of broccoli made my stomach turn, and the texture of asparagus sent my gag reflexes into full gear. My parents were beside themselves because no amount of punishment—even the threat of *no dessert*—could get me to take a bite. Finally, out of desperation, my mom and dad tried a different approach: for every serving of veggies I ate, they would give me a quarter. Suddenly, those carrots weren't looking so gross. To an eight-year-old, that modest but steady stream of revenue meant that I'd have enough cash each week to buy a whole pack of baseball cards and countless two-cent Tootsie Rolls from a local market.

Only a few months after the introduction of the Backer

Vegetable-Consumption Incentives Program, not only was I eating my vegetables at every meal, but I was actually starting to enjoy them—and I was making a profit to boot! I was happy, my parents were happy, and even my two older sisters were happy, because they no longer had to witness any more dinnertime battles. (Looking back, I'm sure they were also upset they weren't getting paid to eat their vegetables too!)

This is a simple but powerful example of how rewards can move people to do the right thing, especially when people have no intrinsic motivation. Because the goal of all policies is to deter or encourage a certain type of behavior (in my case, eating vegetables), it's important for policymakers to understand what motivates people to comply with rules and mandates.

Historically, people on the political left have favored federal regulations to get the job done. Such regulations are enforced by punishments, such as fines and legal ramifications. The threat of these figurative *sticks* is meant to prod individuals and businesses into compliance because, let's face it, no one wants to get punished or penalized. Regulations certainly have their place in climate change policy, as you will soon see. Yet, just as the threat of *no dessert* didn't make the younger me budge on eating my vegetables, punishment can go only so far in getting people to do something, even if it's for the general good.

People on the political right, on the other hand, typically lean toward a hands-off approach, believing that innovation and enterprise will nearly always find the best solutions and practices when they are not interfered with. Evidence of the efficacy of a hands-off approach in new businesses is abundant indeed. When company owners are driven purely by free-market competition to come up with new products and solutions, they very often do.

Incentives, if you haven't already figured it out, fall somewhere on the spectrum between command-and-control regulations and a laissez-faire (i.e., hands-off) attitude. Both liberals and conservatives have advocated for an incentives-based approach at different times, depending on who benefits. Just as it took a proverbial carrot to get me to eat actual carrots, most people respond positively to rewards when asked to change a behavior they don't feel inherently inclined to change. Case in point: by the time I got to high school, I began enjoying eating vegetables, not only because their taste grew on me, but also because as an athlete, I was motivated to keep myself healthier. At this point, my parents took a hands-off approach, not only because I was too old to pay off with a quarter, but also because it was no longer needed.

Without incentives, none of the energy sources we use today would have become as inexpensive, accessible, or clean as they are. Oil and gas became as ubiquitous as they did because federal, state, and local governments continually used incentives to generate competition among power companies to encourage better efficiency. The same has been true for renewables. Likewise, throughout the decades, many farmers have survived because of government subsidies and tax breaks that stabilized their incomes. In an open and competitive economy, examples abound of companies that have been able to solve seemingly unsolvable problems and even change the face of whole industries because of a baked-in drive to be the best at what they do. Sometimes, it's the sheer demand of an industry's consumers that motivates one company to outperform all the others, as was the case with the world's first mass-produced hybrid car model, the Prius.

Many people don't know that Toyota never set out to create a hybrid car—even though that's exactly what happened in 1993, when the Executive Vice President of Research and Development at Toyota commissioned a committee of designers and engineers to come up with a car for the twenty-first century. Following consumers' demand for roomy, fuel-efficient, environmentally friendly cars that still had all of the features of traditional modern cars, the small team put their noses to the grindstone and came up with a unique two-engine design. They weren't under the gun trying to follow a sweeping government regulation, nor were they driven by a financial incentive for low-emissions vehicles. The intrinsic motivation behind the Prius was the desire to innovate.

By 2007, driving a hybrid was the cool thing to do, and, thanks to endorsements by celebrities such as Leonardo DiCaprio, the Prius had developed a cachet great enough that it outsold all other hybrid models on the market. Drivers could feel good about using less gasoline, even though the savings in fuel didn't make up for the cost of the car itself. And although they collected no tax rebates for their purchase, Prius owners were satisfied just knowing that they were environmental trendsetters. It was a win-win for the car manufacturer and also for the new climate change–conscious consumer. Most notably, the whole thing was largely pulled off with a hands-off approach.

When Hands-On Works

You may remember the term *tragedy of the commons* from your high school economics class. It refers to individuals' tendency to make decisions based on their own personal needs, availing themselves of a shared resource, regardless of the negative impact that doing so may have on others. Climate change has been called the great *tragedy of the commons* of modern times. While this may sound extreme, to an extent, it is accurate. Behavior as simple as the overconsumption of coffee has led to habitat loss that has endangered 60 percent of coffee bean plant species around the world. Likewise, overfishing has pushed many species, such as the Pacific bluefin tuna, toward extinction in recent years. Fashion overproduction has created so much product surplus that luxury brands have been known to burn millions of dollars' worth of a season's leftovers just to avoid offering a discount on unsold wares. The list of these types of *overs* in human behavior is endless.

Clearly, many communities, when left to their own devices, have failed and will continue to fail to control behavior leading to the harmful and continuous destruction of the world's atmosphere and oceans. As I've already discussed, every post–Industrial Age society that has risen to great heights has also seen a great rise in greenhouse gas (GHG) emissions. This is an incontestable fact. However, some on the political left take the idea too far, pegging capitalism and the free market as climate change's worst culprit, as when Nicholas Stern, one of the earliest climate change activists, called global warming "the greatest market failure of all time."

While I disagree with this oversimplification, an unmistakable need does exist to correct some of the damaging effects that industrialization has had on our planet. Otherwise, of course, I would not do the work that I do; nor would I be writing this book. However, the emergency measures many in Washington and around the world are proposing to take have the potential to create more harm than good. It all goes back to the discussion of trade-offs and taking a nuanced approach to problem-solving that requires patience, collaboration, and innovation.

The enforcement of fines to pay for harm is a four-hundred-year-old idea in Anglo-American law that goes back to when British pig farmers were told to stop stinking up their neighborhoods. According to Professor Dan Esty, it's a simple rule: "Either stop the harm, or pay for the harm. [It] changes behavior in fundamental ways." Over the past several decades, governments, including ours, have sought to mitigate environmentally destructive behavior by instituting various regulations across a multitude of industries. Two basic types of traditional regulatory approaches include 1) a technology or design standard, which mandates specific technologies or production processes that polluters must use to meet an emissions standard, and 2) a performance-based standard, which also requires that polluters meet an emissions standard, but allows the polluters to choose any available method to meet that standard.

Both types of regulations come in handy under critical circumstances, such as war, a pandemic, or natural disasters. When large swaths of a population's health is at stake, such as with the discovery of asbestos's lethal effects, the threat of punishment is often the way to go to make sure swift action is taken. Under the right conditions, regulatory approaches can even produce immeasurably positive results that last for decades.

America once had some of the worst air and water pollution in the world, with smog encompassing most major cities. Even without today's handy smartphones that can track the Air Quality Index (AQI) at any given moment, any person living in Los Angeles in the 1960s could tell you that the air they were breathing was unhealthy. (Hint: if you can "see" air, it's most likely not just air.) In response to these dire circumstances, Republican President Richard Nixon signed the Clean Air Act of 1970, which became one of modern America's most consequential laws, reducing air pollution in the United States by 66.9 percent to date.

Most notably, the act was a fully bipartisan effort in the midst of substantial political upheaval, much like how many would describe America's milieu today. Just two years after the assassinations of Martin Luther King Jr. and Senator Robert F. Kennedy, and with the US's controversial involvement in the Vietnam War, the American public had lost confidence in the government's ability to solve important problems. The Clean Air Act was instrumental in restoring that confidence.

Times of division often become fertile ground for such unlikely collaborative successes, and I believe 2024 and beyond are no different. Yet, unlike the current proposed GND, the Clean Air Act wasn't tethered to any other political reforms. Rather, it was singularly focused on restoring the environment, setting standards that the nation's air must meet, based solely on what was best for Americans' health. And with these performance-based standards, businesses and industries were given autonomy to decide how they would meet them.

The wild success of the Clean Air Act can be attributed not as much to genius planning as to a sheer will to meet mutual goals. Considerations of cost and technological feasibility were secondary to these goals; the drafters of the bill were determined that the country would meet them, even though the path toward them was still uncharted. Republican senator Howard Baker of Tennessee, who later became Senate Majority Leader and President Reagan's chief of staff, was one of the greatest driving forces behind the act's aggressive performance-based standards. He insisted on requiring companies to cut pollution faster than existing technology allowed, thereby forcing them to rely most heavily on the strength of innovation, much like the Prius success story I just mentioned.

Of course, as with all regulations, the Clean Air Act was at first met with resistance from certain sectors that felt unfairly targeted and burdened by restrictions. Carmakers, for example, protested that the law's call to cut pollution by 90 percent was unreasonable. However, in the end, they rose to the occasion and achieved this goal—and beyond. Thanks to the act's stringent standards, today's cars are a full 99 percent cleaner than those produced before 1970.

Also instrumental to the success of the Clean Air Act was the fact that, generally speaking, the public consensus was that something needed to be done to mitigate the poor health conditions, which were reaching dire proportions. Among the earliest achievements of the act was the removal of lead from gasoline—and subsequently from the air—resulting in a reduction of blood lead levels that prevented a spike that could cause a child to lose two to four IQ points. Because the future of the next generation was a concern for most parents, including even car manufacturers and factory owners, any initial resistance toward the stringent standards gave way to compliance for the mere fact that it improved the quality of life for almost all Americans.

Another one of the most successful and effective environmental treaties ever negotiated and implemented in the world was the 1987 Montreal Protocol. With an unprecedented level of cooperation and commitment shown by the international community, it was the first treaty of the United Nations to receive universal ratification—by 197 countries. Much of the negotiation for this plan, which aimed to ban the global production and use of ozone-damaging chemicals, occurred in small, informal groups. This dynamic enabled a genuine exchange of views among scientists and policymakers that was relatively unhampered by hidden agendas.

At the time, the science around GHG and the ozone layer was not definitive; therefore, a highly flexible policy was called for, which could increase or decrease controls as time passed and the science became clearer. This flexibility meant the protocol could be amended to include stricter measures if needed, such as augmenting the list of ozone-depleting substances, or a total phase-out of substances, rather than a partial phase-out. Under the protocol, for example, 142 developing countries were able to meet the 100 percent phase-out mark for chlorofluorocarbons (CFCs), halons, and other ozone-depleting substances (ODS) by 2010.

Besides built-in flexibility, the protocol's clear articulation of targeted chemicals and sectors allowed governments to prioritize early the main sectors—namely, refrigeration and air conditioning. Good timing was also a factor in the plan's success, since the CFCs that flooded these two industries were outdated and ready to be phased out in exchange for newer, reasonably priced technologies that would benefit both the environment and industry.

A final reason for the protocol's successful implementation was its compliance procedure, which was designed from the outset as non-punitive, meaning no *sticks*, so to speak. It prioritized helping wayward and developing countries, letting them work with a UN agency to prepare an action plan to get themselves back into compliance. The protocol also offered incremental funding for developing countries to help them meet their compliance targets through their Multilateral Fund, available for short-term projects.

When Regulations Backfire

Growing up, my parents would always remind me to shut off the lights whenever I left a room because not to do so would be a waste of money. They didn't punish me if I forgot to flip the switch, nor did they pay me to remember to do so. Intrinsically, I shared their frugality and didn't want to waste money either. Therefore, it was an easy rule to follow.

However, truth be told, not all of my parents' rules were. You may relate to the common bedtime rule of keeping your phone downstairs that many teenagers have to abide by. I know that my parents' reasoning behind this rule was, of course, well-meaning: they wanted me to get a good night's sleep. However, as a young teenager, I didn't share this concern, and, therefore, didn't feel the need to comply with the rule. Instead, I would wait until my parents had gone to bed and then sneak downstairs to secretly reclaim my phone.

The rule ended up backfiring because now I was staying up much later than usual to wait to use my phone, and then, once I had it, my adolescent stick-it-to-them attitude would kick in and I stayed on it even longer than I normally would have had I had free rein to begin with. (Sorry, Mom and Dad, for the public confession!) The point I am making is one of basic human psychology: when regulations are too restrictive, people will go to great lengths to find a loophole in order not to comply.

Besides resistance from certain sectors or individuals, one significant challenge that all regulations face is longevity, a quality connected to questions of when to update policies or terminate them altogether. The Clean Air Act has stood the test of time because of the future-proofing that its drafters originally built into it. For example, among the many systems of accountability embedded in the act is the requirement that the Environmental Protection Agency (EPA) review air-quality standards every five years and update them in accordance with the latest evidence.

Some regulations, however, are not nearly as agile as the Clean Air Act, partly because of intrinsic flaws within their initial structures that do not account for an ever-evolving society or for technological advances. Other failed federal regulations address issues that would have been much more effectively handled at a local or state level. Still other regulations falter simply because a hands-off or incentives-based approach would have been the best choice in the first place.

The Endangered Species Act of 1973 (ESA) is an example of a policy that no longer works because of its outdated regulatory structure, and while it doesn't directly relate to climate change, it does hold similar stakes for those concerned about the environment. The goal of the act was, of course, to raise awareness of wild species whose numbers are dangerously declining and to provide measures that would protect these species from extinction. The metric for the act's success, therefore, would be an ever-shrinking list of endangered species. However, since its inception, the new listings of endangered species have far outpaced the delistings, a disparity that has caused much controversy over the act's efficacy.

Many factors come into play when determining whether a species remains on the list. First, delisting requires an evaluation of the regulatory protections that will remain after the removal of the species from the protected list. The details of such a study would vary from region to region, depending on a number of factors, and, therefore, would take years of effort. Second, a lack of effective protection against the causes of a species' decline means most species will remain on the list forever. Finally, the negative consequences of erroneous delisting generally exceed those of erroneous retention on the protected list, and therefore, many environmentalists advocate against delisting in general.

Another controversial aspect of the ESA is that it includes what is called the *take clause*, which makes it unlawful for any person "to harass, harm, pursue, hunt, shoot, wound, kill, trap, capture, or collect, or to attempt to engage in any such conduct" any listed species. Of course, this clause is paramount to a species' chances of survival and seemed perfectly sensible at the time it was written; however, the government's subsequent addendum to include the *taking* of habitat that harbors—or could harbor— endangered species has led to adverse consequences for the species it was designed to protect.

The extension of the government's jurisdiction to include private property is where the regulation begins to meet with serious resistance, as private property owners have very little motivation to abide by it and a lot more motivation to protect their property. The choice between having the federal government begin to monitor their land or just letting a rare fox die rather than keeping it alive seems like a no-brainer to many farmers, foresters, and homesteaders. With no incentive to go out of their way to save that fox, they simply won't. In fact, many landowners have, unfortunately, participated in a practice that has become known as the *Three S's—shoot, shovel, and shut up—*just to avoid government interference in their everyday lives.

Private landowners have found additional ways to circumvent the ESA's restrictions on their land. Because the federal government limits the amount of timber harvesting in areas occupied by endangered species, private foresters have been notorious for preemptively harvesting their trees to deter endangered species from inhabiting their property in the first place, thereby diminishing the species' available habitat and endangering them even further. It's hard not to wonder in cases like this whether a hands-off approach would have been better. After all, private landowners are naturally inclined to protect their forests, not because the government has threatened them with penalties, but simply because they care about their property.

The True Costs of Regulation

In times of crisis—again, by that I mean such circumstances as war, a public health outbreak, floods, and terrorist attacks—regulations are important to save lives, regain national security, or restore equilibrium. However, when it comes to climate change, strict regulations are not always good for the environment. Bureaucratic red tape unnecessarily slows down projects and drives up costs, hurting investments, and actually disincentivizes progress.

The National Environmental Policy Act (NEPA) is a law that requires all federal infrastructure and energy projects to undergo an environmental assessment and Environmental Impact Statement (EIS) for the project to be completed. Originally, NEPA was intended to protect the environment, but over the years it's become a heavy bureaucratic burden that has slowed down progress and generated unnecessary expenses in the green energy movement. NEPA's review process takes an average of seven years to complete, and adds an average estimated cost of $4.2 million per project. Yikes!

The evidence shows that NEPA is an outdated environmental law impeding clean energy progress from originating on American soil. Instead of trying to get through its obstacle course, many manufacturers opt out of updating their facilities to meet NEPA standards and instead import products from China or other countries that produce even higher emissions. In 2020, the Council on Environmental Quality passed a final ruling to modernize NEPA by streamlining the EIS and reducing its timeline from seven years to no more than one year. That's progress, but not enough.

One surefire way to make the approval process for infrastructure projects more efficient is to keep a majority of regulations at a state and local level to begin with. Only those policymakers who are familiar with each state's specific infrastructure, needs, and limitations can know what's truly best for that state, and what is a plausible ask for that particular locality. It goes back to the question of intrinsic motivation. If state and local officials were given more authority to approve projects that directly improved their jurisdictions, I guarantee that we would see a significant increase in the rate at which these projects were approved.

The construction of a local energy facility in remote Iowa, for example, which would generate three thousand new jobs within the region, will likely be approved a lot more quickly by a county clerk in Des Moines than it would having to go through Congress. Likewise, an offshore wind farm off the coast of Long Island would receive much more immediate support from the New York governor than it would by a federal official sitting hundreds of miles away at a desk in DC. While this may be true, unfortunately, individual states and counties are almost never consulted during the regulatory process. In fact, most likely, the staff that puts together regulatory suggestions has never even visited the states in which they will be hardest to implement.

The Inflation Reduction Act of 2022 (IRA) marks the most significant government investment in fighting climate change in US history, with the goal to reduce emissions by 40 percent by the year 2050. But that's only if the money is actually used as it was intended and if the projects it is hoping to support are actually built. Unfortunately, considering the costly delays that heavy regulations incur, those are huge *ifs*. Many projects face bankruptcy before they even reach the approval stage, as you will see in just a moment when I discuss nuclear energy.

While the IRA is a big boost to our country's efforts to lead in the global clean energy movement, its big-picture concerns may in fact impede actual big-growth progress. Again, it's the revolution versus evolution mindset I introduced in Chapter One that touts massive-scale growth, yet does very little to move projects along at the local and state levels where true innovation happens.

It's easy to understand why federal officials don't feel the need to expedite a project that will provide clean energy for only 2 percent of a state's population or approve construction of a facility that will employ only 150 people over the course of two years. Yet, added up, the aggregate impact of such projects is exponential and has the potential to be expedited at much faster rates.

The obstacles and delays worsen when the government props up certain industries based on which ones they've determined to be *winners* in advance of free-market outcomes. I've talked to countless CEOs who contend that a majority of their power would come from nuclear or wind power if they were actually allowed to build the infrastructure to support these types of clean energy. Instead, corporate leaders understandably choose the path of least resistance when it comes to clean energy, which historically has been solar energy, mainly because its permitting process is comparatively simple and easy.

But fundamental problems abound when we pick winners and losers, not the least of which is that sometimes the winners turn out to be losers in the end. Solar and wind energy, though subject to less bureaucratic hamstringing than nuclear and other clean energy, are riddled with their own set of moral and security trade-offs that we must now face.

The Need for Permit Reform

Unlike getting a driver's permit, the permitting process for renewable energy projects is multilayered and often requires approval from local, state, interstate, and federal authorities. Local communities may oppose projects such as solar panel fields or wind turbines, which create unwanted noise, potentially endanger wildlife, or obstruct their views. These concerns could be voiced through public comment processes, litigation, or advocacy; many towns and counties are imposing suspensions on renewable development altogether. In response, states may institute higher environmental review burdens or block unpopular projects.

Federal authorities may take years to grant the full range of permits a project requires until finally it becomes outdated or underfunded. The exact type and number of permits for a particular project depend on its size, geography, technology, and jurisdiction. Often projects are delayed at any given stage of the process and require multiple consultations and assessments to go further. State environmental laws can also stall projects, through the review process or their use of litigation by project opponents. Most notably, the California Environmental Quality Act (CEQA), which resembles and sometimes goes beyond NEPA, has frequently been wielded by local opponents for various reasons to block climate-friendly projects such as more bike lanes, bus rapid transit, and wind turbines.

In early 2023, I toured Culdesac Tempe in Arizona, America's first car-free apartment community, where residents receive ongoing incentives to ride bikes and use Lyft rideshares and Bird scooters. They are also given complimentary Platinum Passes for unlimited light rail, streetcar, and bus rides whenever they wish to venture into Phoenix's urban sprawl. Located among sixteen acres of curated courtyards, shops, and gourmet eateries, each apartment unit also can be reshaped into different layouts, meeting the desires of its residents.

Sound too good to be true? It was when the Culdesac development company thought about setting its inaugural location in California, a state whose climate action ideology would seem to embrace such a community. Ironically, however, California is also notorious for its costly, time-consuming regulations and permit-approval processes. According to the people I spoke to at Culdesac, the project simply would have been a nightmare to even start there, much less see to fruition.

In my opinion, the dichotomy between California's aggressive climate agenda that calls for a ban on all gas-powered vehicles by 2035 and the state's crippling regulations that would prohibit a carless community is backward at best and pure dishonesty at worst. It leads me to question what the true motives of states like California are in the climate action game. In contrast, Arizona, a traditionally red state, has embraced this unique, environmentally forward opportunity to become a forerunner in carless communities, without conflating the project with any other environmental agendas.

Clean Energy Hamstringing

Often renewable energy projects can encounter challenges due to their interconnectivity with the electric grid. The queue system that most grid operators use, for example, has created a backlog of applications that has nearly doubled the wait time for implementation. Federal permitting of renewable energy projects, more than local, state, or regional challenges, can take months or even years for approval. Under Section 404 of the Clean Water Act, for example, a key permit required for energy contraction has been known to take an average of four hundred days.

Offshore wind (OSW) is among the most backlogged renewable energy projects around, mostly due to an insufficient amount of personnel to keep up with the recent surge in requests. At the time of this writing, the US has only forty-two megawatts of OSW capacity installed. Compare this to China, which installed seventeen gigawatts of OSW capacity last year alone. While countries like China (and European nations) are building offshore wind, the US government is spending over ten years to approve the permits for offshore wind projects.

Likewise, America's nuclear energy sector has some of the toughest restrictions and permitting regulations in the world. Last year, I toured the Alvin W. Vogtle Electric Generating Plant, also known as Plant Vogtle, in Waynesboro, Georgia, where the most common complaint I heard among employees and leadership was about the exponentially higher costs and amount of time that unnecessary government regulations slapped on to each project.

Indeed, it's no secret that the Nuclear Regulatory Commission's (NRC) heavy-handed regulatory approach has made it nearly impossible to build any new nuclear power plants in thirty years. In 2009, the Georgia Public Service Commission approved the construction of two new reactors at Plant Vogtle; however, just three months later, the NRC decided to impose the aircraft impact rule, which requires all new reactors to be able to withstand direct impact from a large aircraft.

The fact that the rule was not applied to existing plants or even to plants that already had obtained construction permits arguably demonstrates its arbitrariness. In addition, the rule's adoption led to a protracted and financially burdensome period of further redesigns that failed repeatedly to pass the tests of conflicting standards. For example, one particular design that was able to withstand the impact of a jetliner was no longer flexible enough to withstand an earthquake. Finally, two and a half years later, the final design certification was approved. By this time, however, the project's budget was overspent and certain deadlines were missed, and ultimately, the entire endeavor came to a halt.

The NRC's prescriptive regulatory approach stifles innovation because it follows the first type of regulation I described at the beginning of the chapter: it mandates specific safety features and designs rather than setting safety standards that allow companies to experiment with the most effective design to submit for approval. In addition, the application of an outdated regulatory framework to a new generation of reactor technologies has been devastating to the scalability of US nuclear energy development. Instead, the licensing process designed by the NRC should be performance-based and technology-inclusive, giving various reactor designs the opportunity to compete for cost-effectiveness and safety within a light-touch regulatory framework.

Other forms of renewable energy have become hamstrung by federal regulation, including geothermal energy. This little-talked-about energy works by harnessing the heat originating at the Earth's core as hot water flows up through porous rock, picking up heat energy along the way that turns into steam. If you've ever seen a geyser erupt or taken a soak in a hot spring, you've experienced this naturally occurring phenomenon firsthand. Capturing this energy typically works by drilling through the rock, allowing steam to travel up into a man-made well where it can then power a turbine to produce electricity. A second well, called an *injection well*, is used to produce a continuous cycling of energy by putting water back into the system.

Although the federal government has deemed only a limited number of sites suitable for traditional geothermal energy generation, next-generation technologies, such as horizontal drilling and hydraulic fracturing, could eventually allow us to drill and capture geothermal energy just about anywhere. A major challenge that remains for many states, however, is the permitting process to drill wells and produce geothermal energy on federal lands, which account for about 90 percent of known sites with geothermal potential. Although the Energy Policy Act of 2005 streamlined the permitting process for oil and gas wells, that provision does not include geothermal wells. Instead, any geothermal energy exploration or development must go through environmental assessment under NEPA, resulting in an average development timeline of eight years.

The bureaucracy surrounding the permitting process for nature-based emissions reduction practices is also far too restrictive. In the following chapters, I'll discuss in detail the clear benefits of forest management practices such as prescribed burns in lowering greenhouse gas emissions. Fortunately, lawmakers are already catching on to these benefits and drafting several bipartisan bills that aim to reduce or eliminate the red tape hampering such projects. For example, both the Root and Stem Project Reauthorization Act, co-sponsored by Senators Dianne Feinstein (D., CA) and Steve Daines (R., MT), and the Save Our Sequoias Act, introduced by Representative Scott Peters (D., CA), seek to simplify and speed up the regulatory process that governs forest treatments.

These protracted timelines have very real consequences, especially when it comes to environmental protection projects such as forest management. One forest-thinning project in Northern California, for example, recently was held up for more than a decade by unnecessary red tape created by activists who wanted to protect endangered spotted owls. As the project was paused, a wildfire tragically burned the owl habitat to the ground and contributed to the 130 million tons of carbon dioxide released by forest fires that year, the equivalent of about a year's worth of pollution from twenty-five million cars. More broadly than just California, the well-intentioned—but harmful—desire for a hands-off approach to protect spotted owls has backfired immensely. Environmental activists prevented forest management, setting the stage for massive wildfires, and 75 percent of spotted owl habitats has burned as a result. All of this could have been prevented via forest management.

It's clear permitting reform is crucial to avoid bottlenecks in clean energy development. The federal government should step in only when projects involve territory across more than one state, such as a river; when projects impact the nation's health, such as with the carbon emissions from forest fires; or when one state's actions affect an entire industry, such as California's stricter tailpipe emissions regulations, which forced automakers to change their designs and raise prices for the entire country.

YOU MAY BE SURPRISED TO KNOW . . .

Another challenge that arises with broad, sweeping regulations is the possibility of reinterpretation years later to accommodate different political priorities and agendas. Over the past half century, the Clean Water Act of 1972 has brought our waters back to life—turning rivers and lakes from dumping grounds into productive, healthy waterways again. It keeps 700 billion pounds of pollutants out of our waters annually, has slowed the rate of wetland loss, and has doubled the number of waters that are safe for fishing and swimming. Levels of metals, like lead, in our rivers have declined dramatically.

For three decades, both the courts and the agencies responsible for administering the Clean Water Act interpreted this phrase to broadly encompass all streams, wetlands, rivers, lakes, and coastal waters. However, in 2020 the Trump administration repealed the 2015 rule and replaced it with an extremely narrow definition of "waters of the United States" (WOTUS) that removed protections from roughly half of the nation's streams and wetlands.

Fortunately, the 2020 rule was swiftly struck down in court, with the judge citing its "fundamental, substantive flaws" and the risk of "serious environmental harm." It became clear this type of narrow interpretation of WOTUS would have a devastating effect on the nation. In desert regions, nearly all streams were left unprotected, and several drinking water supplies had already become contaminated. In Georgia, the rule allowed the construction of a mine on 400 acres of wetlands immediately adjacent to Georgia's Okefenokee National Wildlife Refuge. And tens of millions of Americans who live in coastal counties were put at greater risk from hurricanes because coastlines were left unprotected.

Since no state-level regulations were in place to mitigate these deleterious effects, regions across the country suffered needlessly.

Not All Carrots Are Equal

At first glance, the dynamics of regulations and incentives seem simple: regulation has to do with punishment, and incentives are all about rewards. And while often people respond much better to rewards, once we begin delving into the question of who is benefiting and who is being penalized, the conversation starts to take on a new set of twists and turns. Add to it the question of equanimity, agility, and longevity, and the subject of incentives grows even more complex. I will say it again: it's all about trade-offs.

Subsidies are forms of financial government support for activities believed to be environmentally friendly. Examples of subsidies include grants, low-interest loans, favorable tax treatment, and procurement mandates. Subsidies have been used for a wide variety of purposes, including brownfield development after a hazardous substance contamination, agricultural grants for erosion control, low-interest loans for small farmers, grants for land conservation, and tax rebates for the use of renewable energy.

Incentives can motivate more companies and consumers to participate in environmentally beneficial behavior. More participants means more competition, which, in turn, means more innovation. Climate policies such as tax credits, subsidies, and other incentives matter most when they catalyze a cycle of higher demand that leads to more innovation, learning-by-doing, and economies of scale that lower costs and further boost demand. In solar power, for instance, the cost of solar panels fell 97 percent between 1980 and 2012. These advances were spurred by the promise of demand that early government incentives made possible.

When electric vehicles first came onto the mainstream market some fifteen years ago, my dad's cousin was *that cool relative* who could pull up to a family dinner in one of them. He'd decided to make a purchase to protect the planet and simultaneously take advantage of the tax rebate the government was offering all EV purchasers. It seemed to be paying off. Back then, the range of distance before having to recharge an EV was a measly thirty miles, a fact that didn't have much bearing on his decision since he lived on Bainbridge Island near Seattle and didn't have to drive more than ten or twelve miles per day on his round-trip commute to work. For most of the rest of my family of Wisconsinites, however, who often drove twenty-five miles one way to work, EVs and hybrids made no sense at the time.

Since those early EV days not so long ago, technology has advanced significantly, enabling EVs to go considerably farther compared to the distance they used to, making them more appealing to rural consumers who have to drive long distances. Yet this doesn't change the fact that they're often still double, even triple the cost of cars with combustion engines.

Additionally, once you read the fine print, even the tax rebate is not as helpful as the government and car manufacturers would have you think. Car buyers must earn an annual salary of at least $66,000 and have no other significant tax credits elsewhere in order to owe enough taxes to get the full benefit of the rebate. For this reason, nearly 80 percent of the incentives for electric vehicles have been claimed by people making at least $100,000 per year. One way to achieve what's been dubbed *rebate equity* among the middle and lower classes would be to redesign EV rebates to include used models as well.

Besides a lack of equity, the rebate restrictions that make EVs available only to those in higher tax brackets have a secondary consequence that most people concerned about climate change don't even know about: these rebates may actually increase, not lower, overall emissions. Because EVs are mostly purchased by wealthier households, often as secondary cars, they are often driven fewer miles and kept for fewer years.

My point here is not to renounce EVs as a part of our climate action plan. It is merely to say it's not a panacea like we often hear. And that when it comes to incentives, we need to take a closer look at what exactly is being incentivized and how much the effort actually will help. For example, over the past few years, California has accepted over $12 billion of taxpayer subsidies for EVs, even though barely 1 percent of all registered cars in that state are EVs. Meanwhile, all of California's progress in emissions reductions over the past twenty years (from all efforts) was completely wiped out by 2020's wildfire season. That's $12 billion in EV subsidies down the drain. Still, at the time of this writing, Gavin Newsom is planning on banning the sale of gas vehicles by 2035 and instituting a new $10 billion subsidy program for EVs.

Considering all factors, it's clear why electric cars still require subsidies to sell. Norway is the only country where most new cars are electric, only because the sales and registration tax on these vehicles—worth $25,160 a car—was omitted, on top of other tax breaks, such as reduced tolls. In order to cut one ton of CO_2

emissions through the subsidization of electric cars, Norway has to sell one hundred barrels of its oil, which emit forty tons of CO_2. As you can see, Norway's subsidization of EVs has to do with that word I've been talking about for a good part of this chapter: trade-offs.

When starting a business, one of the most important first steps is coming up with the initial capital (via private investors or the government) to fund the project. In the same way, incentives done the right way give grounding for success, and at the same time help a technology to become competitive enough to eventually survive on its own. The problems arise when policymakers lose sight of the original purposes for these carrots and over-incentivize. If a start-up donut shop, for example, starts to go into deficit, yet the bank continues to fund the failing business and bail it out of debt, then the owners will never have a reason to re-think their strategies.

Of course, a bank has an inherent motivation to periodically check on the donut shop to make sure its investment is sound and that the owners will be able to pay back a loan. The government, on the other hand, often continues to dole out taxpayers' dollars to incentivization programs for endeavors that are ultimately doomed to fail. You may remember the 2008 bursting of the largest real estate bubble in American history, which led to a record amount of mortgage foreclosures and an all-out recession. We should heed serious warning from 2008's economic catastrophe when it comes to blindly incentivizing different climate technologies, especially if there's little conclusive evidence of that technology's ability to stand on its own.

Grassroots Approaches Led by Local Communities

Having a balanced mix of hands-off, regulatory, and incentives-based initiatives at local, state, and federal levels is the only way policymakers will be able to address the nuances of various climate change issues as they pertain to individual communities. Just as important, we need to start from the bottom up.

Take the ubiquitously successful composting programs in the Seattle area that have all been led by local communities and government officials who best know how to take advantage of their region's climate and infrastructure. Had the federal government mandated a nationwide composting program instead, states with much drier conditions and more spread-out communities would most likely have failed miserably at achieving these same goals. Additionally, the Seattle composting program offers its services at no cost to its residents, while homeowners still have to pay for regular garbage pickup. Therefore, the more people compost, the less garbage transport they will have to pay for, resulting in significant annual savings.

Over the years, the EPA has pursued a number of non-regulatory approaches that rely on voluntary initiatives to achieve improvements in emissions controls. These programs act as important complements to existing regulation by encouraging polluters to go beyond what is mandated. In the case of clean energy, more incentives are needed to provide the balanced approach I've been talking about. Under the IRA, OSW developers are encouraged to accelerate construction by being granted the most generous incentives to projects that begin building before 2024, with provisions becoming less lucrative over time. The IRA also seeks to spur significant cumulative investment, create an ecosystem of developers, and produce cost declines as the industry moves along the learning curve and finds domestic supply-chain solutions.

A common criticism of command-and-control policies is that firms are encouraged to reduce emissions only to a regulated level and no further. With market incentives, on the other hand, firms will reduce their emissions as long as it is financially valuable for them to do so. This type of motivation helps a company to commit long term to being pro-environment and to continue to innovate in that direction. A market-driven approach also often means cost savings to customers, making this type of incentivization a win-win for the economy and for the environment.

The main disadvantage associated with economic incentives is that they can be inappropriate for dealing with environmental issues that pose equity concerns. Emissions trading programs, therefore, could have the unintended consequence of concentrating pollution in economically disadvantaged areas, also known as *pollution hot-spots*. Point sources of pollution, such as pipes, ditches, and smokestacks, are often located in poorer areas because decision makers know they are unlikely to face opposition from these communities. Communities like these don't have the proper financial or educational resources to push back. So, while the overall air quality of an area may improve because of incentives, the specific locations where lower-income communities share industrial space most likely will not.

Trading programs are cost-effective approaches to environmental protection because firms are granted the flexibility to either reduce their own emissions or purchase pollution *allowances* from other firms who have reduced below their required level. An example is the US Acid Rain Program, a cap-and-trade system that cost-effectively reduced sulfur dioxide emissions from electric utilities. Other examples include voluntary carbon trading schemes, such as the Chicago Climate Exchange, and nutrients-trading programs (between water-polluting firms and agricultural producers) that aim to reduce excessive loading of fertilizer and pesticides into water bodies. You'll learn a lot more about local initiatives in the second half of this book.

The Critical Role of Incentives for a Cleaner Future

As I've said all along, no silver-bullet solutions exist to climate change or the issues surrounding it. While I'm in favor first of a hands-off approach that allows free-market competition and conservationist values to motivate individuals and communities to make environmentally sound choices, I also believe in the great efficacy of thoughtful incentives.

As we've already seen with solar and wind, early investment is key to driving down costs. The greatest potential in incentives programs such as the IRA lies in their ability to drive down costs of new clean energy technologies to encourage widespread adoption. The most difficult sectors to decarbonize, including agriculture, industrial processes, and aviation, need the biggest boost. Green hydrogen, a promising form of clean energy that I'll discuss later in greater detail, is far more reliable and accessible than solar and wind, yet currently, it is cost-prohibitive for many industries to adopt. That could change, however, with incentives such as the IRA's tax credit of up to $3 per kilogram, which will hopefully increase demand and output and encourage the same learn-by-doing innovation that was needed to catapult solar and wind into the affordable clean energy stratosphere.

In addition to kickstarting market-driven solutions, incentives can and should step in when a regulations-based approach is no longer working. As my simple eat-your-veggies story at the beginning of the chapter demonstrates, when an inherent motivation to do something is lacking, incentives work, especially when regulations don't. As we've seen in the case of the ESA, its *take clause* has backfired and actually further endangered the various species it originally aimed to protect. What if an incentive had been put in its place instead? Consider how much more motivated a farmer or forester would be to do the right thing—in this case, protect the species instead of killing it or destroying its habitat—if they were offered some type of tax break or monetary reward instead of being threatened with property restrictions or fines.

According to Dan Esty, coming up with thoughtful incentives can be a powerful way to gain bipartisan support for climate action: "We've seen that the parties can come together on creating incentives. Just in the past couple of years, there's been a push to try to drive more finance into carbon capture and storage." He cites the bipartisan 45Q provision that offers a tax credit for carbon sequestration, as well as reforestation incentives that both the Left and Right have gotten behind. "Democrats like trees. Republicans like trees. There's a big opportunity in these nature-based solutions that with the right incentives could be expanded."

Finding ways to compensate landowners for their conservation efforts holds great potential for the bottom-up, think-big, start-small approach I've been talking about throughout this book. It is also a way to address the rural communities that are often excluded from the rewards that many incentives offer. Whether it's bird hunters renting prairie potholes to provide more duck habitats, fishermen leasing instream flows to increase salmon habitat, or environmental groups compensating livestock owners for wolf depredation, incentives have the potential to unlock pockets of progress that when added up make a huge difference.

Ted Turner is one of our nation's greatest private conservationists. In the early fall of 2021, I visited his sprawling Flying D Ranch outside of Bozeman, Montana, where I sat down with Brian Yablonski, the CEO of the Property and Environment Research Center (PERC), to discuss why incentives for private landowners are instrumental in helping the fight against climate change. "Most of these issues are not going to find federal solutions at the end of the day," Yablonski told me as we sat only feet away from wild, grazing bison. "Free market environmentalism recognizes that most of these issues are about local knowledge. The people who are going to know best are those who are actually in these forests and valleys. These are the ones we need to empower to make the best decisions for the land."

Public-private partnership is the cornerstone to many efforts that have been working to incentivize conservation efforts. The sale of property rights to organizations such as The Nature Conservancy offers ranchers and other landowners important tax benefits while ensuring that this same land will remain protected. PERC is devoted to expanding efforts like these conservation easements across the country with creative solutions that support private land stewards. By creating funds, for example, that compensate cattle ranchers for livestock lost to wolf predation or disease spread by wild elk, organizations such as The Nature Conservancy have produced far greater results in protecting wildlife than any top-down regulation such as the ESA could ever hope to achieve.

Instead of heading straight to Washington to create a bill intended to protect public land across the nation, granting public land managers more decision-making authority would make more sense. A superintendent at Yellowstone National Park, for example, has far more hands-on knowledge about how best to help a local species of trees than a man or woman inside a DC conference room ever would.

Aldo Leopold, one of the founding fathers of the modern-day environmental movement, once posited, "Conservation will ultimately boil down to rewarding the private landowner who conserves the public interest." The bipartisan Growing Climate Solutions Act that many of my colleagues and I helped pass in 2021 is a perfect example of this dynamic. The act helps farmers, ranchers, and foresters explore emerging voluntary markets that will compensate them for reducing emissions on their property.

Private conservation efforts can and do reap benefits that extend far beyond property boundaries, especially when nearly 70 percent of US land is privately owned. Similarly, supporting local energy innovation is crucial to our climate action plan. As you'll see in the next chapter, this will require us to bring the mining and manufacturing required for renewables production back home to American soil. Similar to the Civil Rights Movement or other important movements throughout history, we can come together to make this happen. Let's explore how.

The 411 from Chapter 2

- Flexible regulations such as the Clean Air Act work when they give businesses and communities decision-making power.

- Climate policy needs to start at local and state levels, resorting last to the federal level.

- Incentives are proven to be most effective in environmental solutions, but still need to be managed and evaluated better, as is especially the case with electric vehicles.

- Innovation does not spring from regulation; it develops from problems that need a smart solution.

- We need serious permitting reform to unlock more clean energy innovation.

3

Unlocking America's Clean Future

Breaking America's Foreign Energy Habit

When Houston Rockets general manager Daryl Morey tweeted his support for Hong Kong protesters in October 2019, China responded by blocking NBA broadcasts for eighteen months. Three years later, on the eve of the Beijing 2022 Winter Olympics, China warned the arriving athletes that speech contrary to Chinese law would be "subject to certain punishment." In October 2022, Boston Celtics center Enes Kanter Freedom spoke out against the Chinese government on Twitter, imploring its government to free Tibet. His outspokenness resulted in a blackout of all Celtics games on

Chinese TV, and the thirty-year-old player was forced into early retirement.

It's no secret that the NBA and other sports enterprises are heavily funded by China, a country of avid sports fans. It's also known that China runs a tight ship when it comes to public criticism of its regime, with noncompliance resulting in serious consequences, as the examples above indicate. Unfortunately, this strong-arming does not end in the private sector. Several US climate-related nonprofits are funded by Chinese companies and would collapse without this support. For this reason, many activist leaders have kept silent about China's totalitarian rule, dirty (carbon-intensive) energy, and human rights violations.

In June of 2020, I personally experienced the fallout of this unwritten rule of remaining silent when I was scheduled to speak at an important event hosted by the Action for the Climate Emergency (formerly known as the Alliance for Climate Education) and just days before the event was informed by email that I was uninvited. One of ACE's staff members had scoured my Twitter account and discovered previous tweets that did not "align with the organization's values." My words had nothing to do with climate change, but one particular tweet linked China to the coronavirus and suggested we rethink our economic and public health reliance on that country, which was enough to cancel me without a proper conversation first. They claimed it was racially insensitive to link COVID to the Chinese government—and offered to provide a coach to train me on being antiracist.

At this point, I was no stranger to criticism when it came to differing political views. However, this extreme reaction to something completely unrelated to the conference felt out of left field and, frankly, scary.

Certainly, China is not the only totalitarian regime with which America has been doing business. Before the war in Ukraine, the US, like much of Europe, relied on Russia as a backup supplier of natural gas. In the wake of the West's strategy to economically isolate Russia by minimizing oil and gasoline imports from that country, the Biden administration turned back to other countries as suppliers. In November 2022, the federal government issued an expanded license to Chevron, the nation's second-largest oil company, allowing it to resume production and importation of oil in Venezuela, yet another dictatorial regime.

The potential amount of oil this new development would actually supply only amounted to about 1 percent of our country's total demand. Still, the re-established negotiations with Venezuela have raised some serious concerns among environmentalists and social activists. Venezuela's oil has been known to be among the most greenhouse-gas-intensive in the world, producing about twice the total emissions per barrel as oil from, say, Saudi Arabia. That factor, paired with the Venezuelan government's horrendous human rights record, tells us that we need to find better alternatives for intermittent backup to renewable power sources.

If the world continues to rely on adversarial countries for dirty energy, any domestic progress in fighting climate change will be canceled out. Russia's leaky, antiquated gas production system has been known to produce methane emissions eight times higher than those from the European Union or US domestic gas and is estimated to accelerate climate change twice as much as the coal it's meant to replace. The world's growing reliance on liquefied natural gas (LNG) only adds to the problem. LNG, by the way, is natural gas that has been cooled down to liquid form and takes up about 1/600th the volume of natural gas in the gaseous state. Although it is odorless, colorless, nontoxic, and non-corrosive, LNG can emit just as much GHG as coal because of the methane that leaks out during processing.

YOU MAY BE SURPRISED TO KNOW . . .

Reliance on non-allied countries also, of course, puts them at an enormous political advantage because energy is vital to human survival. Chinese-Russian collaboration on LNG is already happening inside the Arctic Circle and could signal a mutual geopolitical stronghold for both regimes. Russia has recently built Yamal LNG, the northernmost natural gas facility in the world, of which Beijing owns 30 percent. Stationed across from a military outpost, more than two hundred wells are set to tap into the equivalent of four billion barrels of oil. In 2019, the first Power of Siberia pipeline opened to pump natural gas from the Russian Far East into China. A second similar pipeline is under consideration, which would connect China to Yamal LNG and make it easier for Russia to weather another boycott from European importers.

Yamal LNG is only part of a much larger Belt and Road Initiative, China's global infrastructure development strategy that began in 2013. The Chinese government is also investing in an Arctic free-trade zone and has upgraded rail and road links between Russia and China. Although both countries have always maintained conflicting interests, this basic relationship remains: Russia has energy to sell and China is eager to buy it.

What can the US and Europe do to mitigate our dependence on Russia? Specific policies that would help limit the security and climate costs of Russian gas include a temporary expansion of natural gas in America and allied EU countries while simultaneously increasing the EU's renewable energy production (including nuclear).

Efforts to de-methanize US gas can greatly increase the EU's openness to additional American LNG exports as well. Any climate benefits of low-carbon natural gas may disappear when too much methane is produced in its production, processing, and transport. Very few people I speak to on a daily basis know that methane, the primary component of natural gas, when emitted into the atmosphere, has twenty-eight to thirty-six times the warming power of carbon dioxide. Even though its lifetime is much shorter than that of carbon dioxide, reducing methane released from natural gas is one of the most potent short-term wins in fighting climate change. Federal rules around pipe leak detection and repair are a simple solution to this problem.

Another easy win in methane emissions reduction is to decrease flaring, a harmful shortcut to handling the natural gas produced in oil production. At many well sites that lack the infrastructure or economic incentive to transport this by-product away for sale, the gas is burned, or flared, at the wellhead. While a handful of oil-and-gas states, including California, Colorado, Ohio, Pennsylvania, and Wyoming, have methane emissions regulations in place, other states such as Texas have none. To give you an idea of the extent to which flaring takes place in some states, in Texas in 2019, enough gas was flared to meet the entire state's natural gas demand. A federal requirement to connect oil wells to gas infrastructure before drilling begins would help resolve this issue.

If implemented carefully over time, each of the measures I've just discussed could stabilize energy prices along with lowering emissions. Just as importantly, collaborating on energy sourcing with allied nations would help reduce Russia's leverage to use restrictions in natural gas supplies as a threat against neighboring European countries. The US could play a huge role here by enforcing stringent domestic methane regulations on oil and gas production and coordinating these regulations with those in the EU to increase lower-methane LNG exports to Europe.

The GHG Shell Game

When it comes to taking responsibility, LNG and oil might be the least of our worries. Let's not forget GHG emissions, the basic metric of climate change progress.

Comparing emissions from each country, China is the largest culprit at 27 percent, with the US second at 14 percent. It's easy to jump to the conclusion that the West is simply pushing off its carbon footprint onto the East, where many of our major imports are manufactured. However, if we look at how global production and consumption-based emissions have changed over time, we can start to see a more contoured picture that's not so black and white.

In addition to the commonly reported production-based emissions, climatologists also take into consideration consumption-based emissions. They do this by tracking which goods are traded across the world. Consumption-based emissions include all CO_2 that comes from the production of imported goods and are added to a country's total. Production-based emissions that come from the production of exports shipped and consumed elsewhere, on the other hand, are subtracted from the country's total.

If you think this sounds complicated, you're right—it is. To keep things simple, it's important to bear in mind that consumption-based emissions can't be offshored. This means that if an American buys a smartphone, the carbon emissions that went into making that phone are allocated to the US, no matter where it was manufactured. Even if the phone was made by an American factory that relocated to Indonesia, because it was sold in the US, the manufacturing counts as consumption-based emissions.

When we look at the change in consumption-based emissions versus production-based emissions over the years, we can clearly see that offshoring isn't why China has greater emissions than we do.

Since 2007, US total emissions have been declining. If this were due to offshoring, we'd expect to see consumption-based

Production vs. consumption-based CO₂ emissions, United States

Consumption-based emissions[1] are national emissions that have been adjusted for trade. This measures fossil fuel and industry emissions[2] . Land use change is not included.

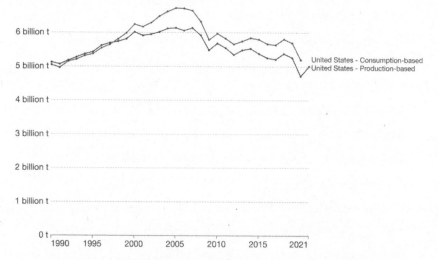

SOURCE: Global Carbon Budget (2022); ourworldindata.org/co2-greenhouse-gas-emissions, CC BY

[1] **Consumption-based emissions:** Consumption-based emissions are national or regional emissions that have been adjusted for trade. They are calculated as domestic (or "production-based") emissions, minus the emissions generated in the production of goods and services that are exported to other countries or regions, plus emissions from the production of goods and services that are imported. Consumption-based emissions = Production-based—Exported + Imported emissions.

[2] **Fossil emissions:** Fossil emissions measure the quantity of CO_2 emitted from the burning of fossil fuels, and directly from industrial processes such as cement and steel production. Fossil CO_2 includes emissions from coal, oil, gas, flaring, cement, steel, and other industrial processes. Fossil emissions do not include land use change, deforestation, soils, or vegetation.

emissions fall a lot less than production-based emissions. But in fact, consumption-based emissions have decreased more, not less, over the past fifteen years. Therefore, our emissions drop is more due to falling emissions consumption (from imports made in China and other countries) than to falling emissions production (from products made in the US). Again, if this is a bit confusing, don't worry. The most important takeaway is that, contrary to popular belief, China cannot blame America's demand for Chinese-made consumer goods for the fact that they have the highest emissions in the world.

If the China emissions problem is not a product of offshoring, then America's efforts to get Chinese manufacturers to reduce emissions through carbon tariffs must be limited in their effectiveness. The fact is, the massive amount of GHG produced each year in China is a problem that only that country can solve. But what is the Chinese government actually doing to solve it? Unfortunately, not enough.

China may lead the world in terms of clean energy, as we will explore in just a minute, but the country still consumes nearly 500 percent the amount of coal that India does, and nearly 600 percent as much coal as the United States (the second- and third-largest coal consumers in the world, respectively). In 2020, the Xi government pledged to be carbon-neutral by 2060. Yet, while global coal consumption has dropped more than 4 percent since then, it has actually gone up dramatically in China. In fact, China is building two new coal plants (on average) each week within their borders—and financing many additional plants in other countries.

Despite its promises to peak in GHG emissions by 2026, China's reliance on coal is likely to last for years, even decades to come. In 2021, the country began building a power plant with the capacity of thirty-three gigawatts of coal-based power generation, which is three times more than the rest of the world's coal-generation capacity in total. China's push to build even more coal-fired power plants, with some two-hundred-plus gigawatts coming online over the next few years, is a serious force to contend with if the world is to meet the current global goal to limit temperature increases to 1.5 degrees Celsius.

The question, of course, remains how much these new plants will actually be used on a regular basis and how many will be reserved for emergencies. An unexpected surge in energy demand (during a heat wave or extreme cold weather), a dip in available solar and wind energy sources, or a sudden disturbance of natural gas imports (as much of Europe has experienced due to the war in Ukraine) are all good reasons to have a reliable backup energy source such as coal.

Unlike China, much of the world, because of limited resources and infrastructure, does not have as many choices for energy. And often those who need the most help are also the ones who are contributing to a large percentage of emissions. In 2021, India, the world's third-largest emitter of GHG, pledged to reach zero emissions by 2070, a target date much later than that of most other countries. Even with this delayed deadline, India's chances for success are slim considering its ambitious plan to get there by generating 50 percent of its electricity from non–fossil fuel sources by 2030.

India also pledged to reduce its emissions intensity—the measure of the amount of GHG emitted per unit of economic activity— by 45 percent by 2030. Even if India were able to make that goal, a fall in emissions intensity does not necessarily mean a reduction in overall emissions. Coal is still a leading source of energy in that country, and its usage is continuing to grow, offsetting any positive impact of cleaner energy sources. The country also failed to sign a deal agreed upon at COP26 to reduce emissions of methane gas, one of the most potent greenhouse gases, of which India is a major emitter.

Because India is now the most populous country in the world, its performance in GHG emissions reduction matters—a lot. It is also a nation that intends to take major steps toward economic development in the coming decades. As I've already discussed, no country in history has managed to grow its economy without an initial surge in emissions. Rather than participating in the global shell game of shuffling emissions responsibility and blame among other countries, the US and other developed nations have a chance to become part of the solution.

If there's anything I've learned over the past several years, it's the idea that every challenge is also an opportunity. Even though at the time of this writing, we are on the eve of a potential global recession, the world is also poised for economic growth in the green energy and climate tech space. In 2021, worldwide employment in the renewable energy sector reached 12.7 million, marking a jump of seven hundred thousand new jobs in just twelve months. Specifically, the demand for solar panels, wind turbines, and lithium batteries has contributed to this surge, inciting a race for dominance in the clean energy market.

China's Growing Clean Energy Dynasty

The world has been harnessing the sun's energy for decades. The first solar panels, also known as photovoltaics (PVs), were mass produced in the 1950s, specifically for space activities. Fast-forward fifty years, and the rate at which the PV domestic market has soared in response to the world's crusade for clean energy is unprecedented. Global solar power–generating capacity, which stood at only 5.5 gigawatts in 2005, jumped to 350.6 gigawatts by the end of 2023, transforming solar panel production into a booming industry. This would be good news, if it weren't for the fact that one country alone now dominates this market by an unfathomable margin—and it isn't the United States.

China currently controls more than 80 percent of all manufacturing critical to the production of solar panels, according to the International Energy Agenda (IEA)—that's right: 80 percent. Over the past two decades, Chinese government policies have aggressively developed economies of scale and innovations that have significantly cut manufacturing costs, making solar panel imports from China a solid 20 percent cheaper than they are anywhere else in the world.

While China seems to have been instrumental in bringing down costs worldwide for solar panels, the reality is not that cut-and-dried. Dependence on China leaves the solar sector vulnerable to disruptions that could be, and have been, triggered by a single event such as a natural disaster, a war, technical failures, or decisions by Chinese lawmakers. The concentration of the industry not only to a single nation but to specific regions within it, has contributed to global bottlenecks that have resulted in unexpected price escalations and extended timelines.

What can America and the rest of the world do to curb China's solar dominance? Because manufacturing costs remain lower in China than in other nations, diversification is unlikely to occur without global policy action. Policymakers need to make supply chain diversification a high priority and maintain open markets to avoid trade barriers among countries outside of China. Doing so will not only create numerous new jobs but also reduce the carbon footprint associated with the production of solar panels, since much of Chinese manufacturing is still fueled by coal.

Besides more aggressive global policy, we need more financial incentives for manufacturing, research and development, and workforce development to expand domestic production beyond panel assembly. The 2021 Bipartisan Infrastructure Law provided significant investments in clean energy infrastructure, including $62 billion for the Department of Energy (DOE) to expand access to energy efficiency, deliver reliable, clean, and affordable power to more Americans, and fund the research and development of new technologies on our own soil. This is progress, but of course, not enough.

Even beyond economics and national security, there is a compelling moral reason why we need to be concerned about China's dominance in the clean energy market. Recent reports have indicated that some of China's biggest solar companies have worked with the Chinese government to absorb workers in the far western region of Xinjiang, a site now notorious for detention and surveillance programs against Muslim Uyghurs and other minority groups. In June 2022, US Customs and Border Protection began enforcing the Uyghur Forced Labor Prevention Act (UFLPA), restricting imports from this region, where torture, indoctrination, and coerced labor have reportedly taken place.

Yet even with these measures in place, human rights groups and trade associations warn that Xinjiang inputs could find their way into solar imports from other countries, given the difficulty of verifying supply chains in China. The inability to track supply chains has also enabled at least four Chinese solar panel companies to avoid US tariff laws by routing their operations through other Southeast Asian countries.

The US and the rest of the world should not have to depend on subsidized, circumvented imports made by forced labor to build our solar energy future. Unfortunately, our long-time dependence on other countries has left us vulnerable and in search of a viable Plan B. And some of the US government's recent restrictions of panels coming from China have led to thousands of closures of American companies that were reliant on China for their supply.

The repercussions of our dependence on China are starting to hit home. A few years ago, my family installed solar panels in our home in Arizona. My dad was hesitant at first, having been told by the solar company that he would start to see a return on his investment only after about six years. He went ahead with the purchase, only to find that a few years later when a panel needed repairs, the solar company had shut down because supplies had been cut off from Chinese trade restrictions. It took a long slew of phone calls and inquiries before my dad was finally able to get the panel fixed. And it has left him questioning to this day, like countless other Americans, whether the investment was really worth the headache and cost.

Still, hope is on the horizon, with several new US companies planning to break ground that would keep the supply chain closer to home. Construction is underway, for example, in upstate New York for an integrated factory complex called Convalt Energy Inc. that could spark demand for American polysilicon, a key raw material in most solar panels, and thus ease dependence on supply chains linked to China. Still, it could take a long time for companies like this to start gathering permits to expand into manufacturing silicon ingots, wafers, and cells, which are all the building blocks of solar panels.

Depending on available incentives, in the next year the US could triple its panel-making capacity. International support is key as well, but is contingent on government backing. South Korea's Hanwha Solutions Corp. has said it is ready to invest billions of dollars to help build a US solar supply chain, from polysilicon to panels, provided Washington creates new incentives. Maxeon Solar Technologies Ltd., which is headquartered in Singapore, has also said it wants to open a US panel factory, again, contingent on government support.

Wind power is another market China has managed to dominate in recent years. Seven of the world's top ten wind turbine manufacturers are Chinese companies. Because China's open topography and long coastline are excellently suited for wind power, all signs point toward wind power continuing to play a central part in China's pursuit of a green energy empire. China's production capacity for wind turbine components is predicted to grow by 42 percent over the next two years.

Still, constraints in grid management and less than optimal turbine models have limited the amount of wind energy generated by China compared to the electricity per unit of installed wind capacity in American counterparts. Another challenge facing renewables in China is the sheer size of the country. Much of its energy production potential is located in the north and northwest, while most of the population is in the east and southeast. To help solve this problem, China is building a massive transmission line capable of transmitting enough electricity for 26.5 million people. While this and similar lines will be used to transport renewable energy, most of the electricity transported through them will still come from fossil fuel power sources.

All That Glitters Is Not Gold: A Downside to Solar

As people who care about the environment, we can all agree on the imperative to reduce global carbon emissions. But the methods of going about this objective can be complicated and require continual reevaluations to determine the next best course of action. As such, policymakers and constituents from both sides of the political spectrum will require a certain level of humility to be able to revisit an original plan and admit to unintended consequences.

When I first began my work with the American Conservation Coalition, I knew very little about what was happening behind the scenes of clean energy production. I automatically assumed that solar, wind, and EVs equaled *good* and coal and fossil fuels equaled *bad*. Ask most people our age and you likely will hear the same line of thinking.

However, as you have just read, when we talk about green solutions, it's important to understand that not everything that sounds good is good. A field of solar panels built in the US might theoretically be an excellent alternative to a coal-generated power plant. But not if those panels were manufactured by residents in detainment camps in the Far East, in a carbon-intensive factory, with minerals dug out of mines by child laborers on a continent being stripped of its natural resources. By the same token, banning fossil fuels sounds like a good idea. But not if doing so creates shortages that in the end make us dependent on foreign energy that's twice as dirty.

Ironically, another set of trade-offs that should be considered when we think about expanding solar power in the US is the damage that giant panel fields are doing to our ecosystems and natural habitats. Nowhere is this intrusion more apparent than in the Riverside East Solar Energy Zone (RESEZ), a 150,000-acre stretch of desert between Los Angeles and Phoenix that is covered in solar panels. This part of the Mojave has been home to Indigenous communities for centuries, as well as the habitat for several endangered species. But the mass construction of solar farms has been depleting the area's water supply and causing air quality and noise pollution issues. It's the renewable energy sector's equivalent of factory farming, and it's become an enormous source of controversy over the past few years.

While RESEZ has the potential to generate twenty-four giga-watts, enough energy to power seven million homes, we need to ask ourselves at what expense. While the loss of dry wash shrubs and trees may seem minor in terms of carbon-capturing capacity, there's more to the story here than meets the eye. The vast root system of this desert vegetation acts as an enormous carbon sink that can store emissions for thousands of years to come. Tear up this land, and we are not only destroying massive carbon-capture potential, but also releasing several millennia's worth of stored carbon back into the atmosphere.

The answer, of course, is not to get rid of solar altogether. That would be swinging the pendulum back entirely the other way, as many activists would have it. Instead, we should focus our resources on developing battery storage that would equip individual households with their own power reserves. This technological advance would release households from the power exchange taking place with grids connected to these mega solar farms.

With many quick-fix solutions that grant short-term wins, we need to have a balanced approach that weighs the long-term consequences. Interestingly, California recently reduced the amount of earnings homeowners can collect through feeding their power back into the grid by about 75 percent. This cut will likely disincentivize new solar panel purchases by 40 percent in the coming year. To the state's credit, the ability to step back and reevaluate a formerly deemed *sure thing* in light of its drawbacks is precisely the common-sense approach to climate action I've been discussing throughout these chapters.

Electric Vehicles: Not a Panacea

Yet another aspect of EVs is often overlooked: namely, where the energy it takes to run them actually comes from. As of now, the electricity that fuels EVs often comes from coal. In order to accommodate the projected manufacturing rates over the next decade, our power grid would need to move to nuclear, hydropower, and other renewable energy for EVs to truly be effective in climate action. As you can see below, the environmental impact of electric vehicles varies substantially based on the energy source used to charge the vehicle—and can be even *worse* than a normal internal combustion vehicle if using power from a fully coal-powered grid.

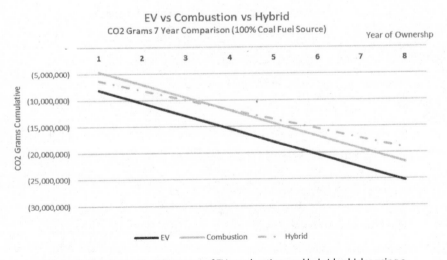

Figure 1. Comparing the CO_2 output of EV, combustion, and hybrid vehicles using a power grid with 100 percent coal

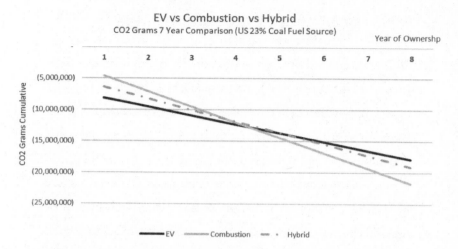

Figure 2. Comparing the CO_2 output of EV, combustion, and hybrid vehicles using a power grid with 23 percent coal (approximately the US share of coal usage on the grid)

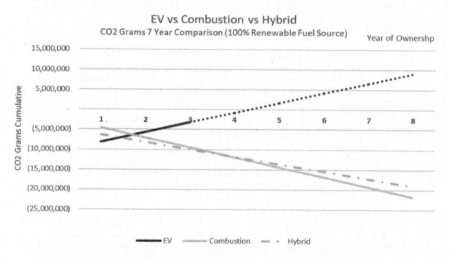

Figure 3. Comparing the CO_2 output of EV, combustion, and hybrid vehicles using a power grid with 100 percent clean energy

In addition to charging, the actual EV batteries themselves pose an unknown threat to the environment. Consider the numbers. Every ton of lithium mined emits nearly seventeen metric tons of CO_2. Each EV battery contains ten kilograms of lithium. In order to actually make the battery, another fifteen metric tons of CO_2 are emitted for every one hundred batteries manufactured. If you add all of this up, that's a whole lot of CO_2 for a product that is supposed to hold the key to net zero. Even under best-case scenarios of lithium mining—with successful recycling and reuse of batteries—we have enough supply of lithium to last us only until 2100. This massive resource constraint has widespread consequences on the supply chain. For reference, with today's technology and electric vehicle resource needs, Toyota claims it can build a whopping ninety hybrids for every electric vehicle built.

In a recent report, Bjørn Lomborg, the former director of the Danish government's Environmental Assessment Institute and current president of the think tank Copenhagen Consensus Center, wrote, "The climate effect of our electric-car efforts in the 2020s will be trivial. If every country achieved its stated ambitious electric-vehicle targets by 2030, the world would save 231 million tons of CO_2 emissions. Plugging these savings into the standard United Nations Climate Panel model, that comes to a reduction of 0.0002 degree Fahrenheit by the end of the century."

The ethical questions about their production should not be overlooked either. Most cobalt, for instance, is excavated in the Democratic Republic of the Congo, where child labor is not uncommon, specifically in dangerous mining practices. And then there is the affordability question. The material prices for batteries this year are more than three times what they were in 2021, and electricity isn't getting cheaper either. None of these challenges are unsolvable in our push for a diversified transportation mix, but we must be up-front about the realities before we dive in too deep.

Making EVs Even Better

China continues to dominate clean energy when it comes to batteries for EVs (as well as mobile devices, electronics, etc.). EVs are predicted to become cost-competitive with conventional internal-combustion engine vehicles by the middle of this decade. Once this important crossover point arrives, EV sales will accelerate. By 2040, pure battery electric vehicles (BEVs) and plug-in hybrid electric vehicles (PHEVs) are predicted to account for the majority of new cars sold and 42 percent of all cars on the road in the US.

Without a doubt, EVs are an important part of a portfolio of clean energy solutions. Built to be efficient, with fewer moving parts than internal combustion engine vehicles, they require far less maintenance than the cars we grew up with and last at least ten years (or 155,000 miles) before the battery has to be changed. EVs use a single battery to power everything from a car's HVAC system to its electronics, to the engine itself, making this component an all-important part of the equation when evaluating trade-offs.

Let's take a closer look. A battery that stops performing above 70 percent of its capacity is only able to be used for short distances, but instead of purchasing an entirely new vehicle, the battery could be exchanged for a new one that could also incorporate innovations and performance improvement. This option is far more cost-effective. The question, then, remains: How can we make batteries even more efficient, and more important, how can we do it domestically?

While companies such as General Motors have their sights on developing a battery with a 250,000-mile capacity as early as this year, American innovation still has a long way to go before it catches up with the EV market demand. Another reason to consider slowing our roll when it comes to pushing EVs is the hard fact that battery production is dominated by foreign manufacturers.

Currently, China is the only country in the world that is able to take lithium from its raw form and process it through to finished batteries without having to rely on imported chemicals, components, or labor. Experts estimate that China's share of the market for lithium-ion batteries is, like solar panels, a whopping 80 percent. Six of the ten largest EV battery producers are based in China, and one of them is responsible for manufacturing three out of every ten EV batteries globally. At this very moment, hundreds of Chinese lithium processing plants, also known as gigafactories, are churning out millions of EV batteries for both their domestic market and foreign carmakers including BMW, Volkswagen, and Tesla.

China's foothold in the EV market began in 2015, when the Chinese government started making huge investments in the complex steps between mining and manufacturing, and an estimated $60 billion in electric vehicle subsidies helped create a market, as well as the battery supply chain to go with it. Even though other countries contain vast amounts of lithium—for instance, Australia is home to almost half of the world's lithium supply—the mineral usually has to be refined and processed elsewhere by China.

China has also taken over giant supplies of foreign lithium (one of the main minerals needed for electric vehicle batteries), including about 40 percent of the 93,000 metric tons of raw lithium mined globally in 2021. China's presence is now on every inhabited continent, with giant deals across South America's *lithium triangle*, a mineral-rich part of the Andes at the junction of Argentina, Bolivia, and Chile, and stakes in Greenbushes, Australia's biggest lithium mine. The same story is true for other rare-earth materials needed for batteries. At the time of this writing, for instance, China owns 70 percent of control of the mining industry in the Democratic Republic of the Congo, home to almost all of the world's cobalt, another critical component of lithium-ion batteries.

To meet the demand for energy storage and EVs by mid-century, the world will need more than twenty times the amount of lithium that was mined in 2021. The good news is Washington appears intent on building a domestic lithium-ion economy, and government incentives for domestic supply-chain development are already underway. For example, the 2022 Inflation Reduction Act (IRA) grants up to $7,500 worth of tax credits to EVs containing batteries assembled in North America and made up of critical minerals sourced domestically. Additionally, the push for *friendshoring*, the practice of building supply chains among democratic allies, is yet another way to minimize China's hold on the market. The US and Europe are already advancing legislation to create a green battery supply chain within Europe, with a focus on recycling lithium.

While these efforts are certainly moving the West in the right direction, the important lithium-processing step between mining and manufacturing still needs to be pried out of China's hands. To help solve this challenge, the US plans to open thirteen new gigafactories by 2025, joined by an additional thirty-five in Europe by 2035. However, in the years it will take to get these lithium-processing plants off the ground, our lithium will still need to be shipped to China and back again to be refined before it can be used in electric cars.

NIMBY and Other Wrong Thinking

New technologies could provide solutions in the interim that would alleviate our dependence on China in other ways. The US and Europe are currently looking for ways to make mining for lithium more economically viable—a way to extract lithium from seawater, for instance—or develop an entirely new type of battery chemistry that does away with the need for lithium altogether. Until that happens, Chinese car manufacturers will still have a huge advantage over European and US competitors. Already, Chinese brands and Chinese-owned European brands are launching EVs in the West that are the cheapest on the market.

As America and other developed nations try to move toward energy independence, the concept of trade-offs that I wrote about earlier becomes even more vital to determine our next best steps. Transitioning to clean energy at the rapid rate that policies such as the GND propose comes at an enormous moral cost as well as a threat to national security. However, if America were to begin mining more domestically, we would be agreeing to the decima-

China's renewables and battery manufacturing dominance

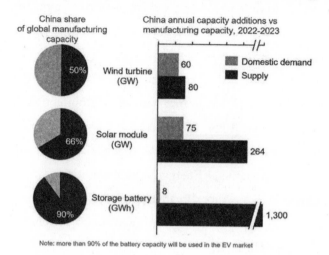

China share of global manufacturing capacity

China annual capacity additions vs manufacturing capacity, 2022-2023

- Domestic demand
- Supply

Wind turbine (GW): 50% — 60 / 80

Solar module (GW): 66% — 75 / 264

Storage battery (GWh): 90% — 8 / 1,300

Note: more than 90% of the battery capacity will be used in the EV market

SOURCE: Wood Mackenzie APAC Power and Renewables Service

tion of beautiful public lands that activists have fought tooth and nail to protect. Let's take a closer look at what's at stake.

The green energy transition has incontestably set off a scramble for resources. And China is not the only country guilty of exploitation and corruption. A quick look at Western history in the Democratic Republic of the Congo—from Belgium's horrific colonial reign to US support for the notoriously brutal and corrupt regime of Mobutu Sese Seko—makes it hard to paint Chinese investors as uniquely evil among the foreign actors operating in that country. Western giants operating in the Congo, such as the Swiss mining and commodity trading company Glencore, have a long and ongoing record of human rights violations and corruption, including bribery and market manipulation, in that region. And as long as US demand for foreign mining and processing continues to contribute to the current supply chain, we are also guilty of the dirty way countries such as China are going about mining for clean energy.

The hard truth is that the world cannot mine and refine the vast amounts of minerals that go into batteries—lithium, nickel, cobalt, manganese, palladium, and others—at anywhere close to the scale for the rapid transition to EVs that policymakers are proposing. The International Energy Agency admitted in a report in 2022 that, to meet its climate change goals, lithium demand must rise by 900 percent by 2030 and 4,000 percent by 2040. That's a lot of lithium.

This is where *not in my backyard* (NIMBY), the great hypocrisy of America's energy revolution narrative, comes in. The same activists who promote wind and solar and EVs as the silver-bullet solutions to climate change also oppose the domestic mining of the lithium and other critical minerals necessary to make them work. As long as this mining is done thousands of miles away on foreign soil, most environmentalists won't even raise an eyebrow. Even if they did, the mining companies (mostly Chinese-owned) would not permit visitors anywhere near these sites, a reality that allows for zero environmental and human rights accountability.

Environmentalists have a reason to be concerned about domestic mining. In addition to damaging precious terrain, the hard-rock-mining industry is a major source of toxic waste. Cyanide, arsenic, mercury, acids, and other substances used to obtain and process ore inevitably seep into aquifers and rivers, sully the land, and are carried long distances by wind. After a century-old copper-mining facility on the Tohono O'odham Nation in Arizona closed in 1999, for example, the Department of Health and Human Services found the mine had added enough arsenic to local drinking water to cause nausea in residents as well as skin, bladder, and lung cancers.

YOU MAY BE SURPRISED TO KNOW . . .

Much of the world's current lithium supply is sourced by capturing it from water via a very slow evaporation process that often takes years to complete. The world's richest lithium resource lies in enormous salt flats in South America's Lithium Triangle region, where it is captured via this evaporative process. But much of the lithium supply is also captured via a hard-rock mining process that is far more impactful on the landscape and environment than the evaporative process. Both forms of lithium capture will have to increase in a very short period of time for EVs and renewables to play their envisioned roles in the energy transition.

The US's General Mining Act of 1872 allowed hard-rock miners to pay minimal fees to stake a claim on land. And unlike coal miners, who pay the federal government an average of about 10 percent of the gross value of what they have produced, hard-rock miners have tendered no royalties on $300 billion worth of minerals extracted from public land over the last century and a half.

The law, which came into effect back when pickaxes were still being used—not today's industrial machinery—also failed to include any protections for land, air, or water. When the fed-

Unlocking America's Clean Future   **83**

eral government first set aside tribal reservations for mining, often in remote deserts and plains, these lands were thought of as worthless. After coveted minerals were identified in these areas, mines proliferated in and around these lands. Today, more than six hundred thousand Native people live within about six miles of hard-rock mines for uranium/vanadium, gold, copper, and lead.

Objections to oil pipelines by tribal nations such as the Standing Rock Sioux in 2016, and to copper-nickel mining near Minnesota's Boundary Waters Canoe Area Wilderness by congressional Democrats in 2022, have raised public awareness of the destruction that mining can cause. However, not many NIMBYists care to admit that mining worse than this is being done all over other parts of the world due to their own aggressive climate change agendas.

Turn on Fox News and you'll hear the argument that if China is increasing its emissions, building more coal plants every day, and using dirty energy to mine the minerals needed for EV batteries and PVs, then why should we even try to do anything about climate change? Switch the channel to MSNBC, and you'll be told that in response to China's lead in the clean energy race, America should sprint ahead with our own mining and solar panel production. Both sides are conveniently leaving out certain aspects of the debate on what to do. Their oversimplification of what's happening in global clean energy impedes progress because it does not take into account all of the trade-offs that are involved.

Moving Toward a Nuclear Future

A popular position among a subset of Americans is that because China and other countries aren't doing enough to reduce their emissions, the United States shouldn't bother to do so either. At the 2019 congressional hearing I participated in, Representative Kathy Castor (D., FL) asked for my take on this position. I answered quite emphatically that this was a faulty approach. I pointed out that never in history has the US looked at a global issue and refused to help solve it simply because of the complacency of other nations.

I also added that we must hold other emitting countries accountable, just as we hold ourselves accountable. Whether we ban specific imports from other countries, place tariffs on those countries' products, or focus more heavily on exporting our own technologies, there are countless steps we must take. I reminded the congresswoman that at that very moment, Hong Kong protesters were waving American flags and singing our national anthem outside of the US embassy as part of their effort to emancipate from China. As I told Ms. Castor, we have always inspired other countries and led in great initiatives in the past. We must continue to do the same with climate change.

However, calling for a revolution is not the answer. Instead, we need to help the rest of the world ease its reliance on unclean energy by finding creative, domestic solutions that form a balanced approach. Realistically, the only way to reduce mining's damage, for example, is to reduce our demand for lithium. Innovative efforts such as expanding mass transit and repairing, reusing, and recycling batteries and other technologies are ideas that need to make swift progress. For now, I'd like to take a quick look at nuclear energy and its promising hope to push the US to the forefront of the clean energy revolution.

Today, Russia continues to be the largest player in the nuclear energy export market by a wide margin. It produces about 35 percent of the world's enriched uranium reactor fuel and has thirty-four overseas reactor units completed or under construction, with nearly $140 billion in foreign orders in 2021. The Ukraine war has scared off some potential clients, but sanctions to this point have largely spared Rosatom, Russia's state-owned nuclear corporation.

An extensive Chinese-Russian nuclear energy collaboration has been an important part of this business. Twenty-seven out of thirty-one reactors that started construction since 2017 are Russian or Chinese designs. In 2021, Russia and China broke ground on their largest-ever joint nuclear energy project: the installation of two new Russian VVER-1200 reactors at the Tianwan plant (the plant's seventh and eighth units), and two more at a new plant in Xudabao.

The Chinese government has approved the construction of six nuclear reactors that will cost an estimated 120 billion yuan ($18.7 billion) and a plan to more than double nuclear power capacity by the end of this decade. Forty of its existing fifty nuclear reactors were constructed in the last decade, and there are plans for at least 150 more reactors in the next fifteen years.

Unlike Russia's nuclear efforts, China's new reactor projects have been mostly domestic, but that is changing. China has begun to sell its own nuclear work abroad. It has exported four reactors to Pakistan and discussed or signed contracts to build nuclear power plants in Argentina, Turkey, the UK, and several African states. This points to a shift in the Sino-Russian relationship in the realm of nuclear power. While China's reliable patronage is a financial boon for Rosatom, China is now largely self-sufficient in its nuclear fuel cycle. It is further weaning itself from Russia by developing indigenous nuclear technology and diversifying its global partnerships.

China's nuclear independence from Russia and its nuclear presence in other countries are worth keeping an eye on. Beyond the economic benefits, nuclear power plant contracts have geopolitical implications. Such plants take years to construct and must be maintained and refueled for decades. This can give nuclear-exporting states substantial political influence in client countries, particularly in developing countries that lack the resources and indigenous expertise to handle plant operations themselves.

Despite Russia and China's ostensible head start, the US still has a big opportunity to take the lead. In the IEA's plan for the world to reach net-zero emissions by 2050, the amount of nuclear power generation has to double between 2020 and 2050. The IEA's plan for nuclear energy includes nuclear power technologies that are not yet available at scale, such as small modular reactors (SMRs), which generate about a third of the energy generation of a conventional nuclear power plant. The lower cost, smaller size, and reduced project risks of these SMRs are likely to improve social acceptance and attract private investment.

At the time of this writing, the US ranks number one in terms of installed nuclear energy capacity, with France and China, respectively, in second and third place by a wide margin, and Russia in fourth. Many new innovations are taking place as nuclear energy gains in popularity and draws attention from investors.

Recently, Microsoft has partnered with TerraPraxis, a nonprofit headquartered in the UK, to develop a software application that will help replace the coal-fired boilers at shuttered power plants with SMRs. TerraPraxis hopes to help automate the design and regulatory processes involved in transitioning coal plants to nuclear power and believes that more digital tools could help some 2,400 coal-fired power plants worldwide transition to nuclear generation. This collaboration is a perfect example of pairing the old and new, and I believe it could be one of the keys to unlocking America's potential as the clean energy leader it needs to be.

Geothermal:
The Dark Horse of Clean Energy

The amount of available geothermal energy beneath our feet is estimated to amount to fifty thousand times more than the global total of oil and gas resources combined. That's a whole lotta heat right below our feet. The benefits of geothermal energy production are obvious. Because heat flowing from the Earth's core is continually replenished by the natural decay of radioactive elements, it is the definition of renewable and 100 percent domestically sourced. Unlike solar and wind farms, geothermal plants can run 24/7 and are not dependent on weather conditions. Geothermal plants also use far less land and water than coal, wind, and solar stations and emit no GHG.

The core of the Earth is a blazing 10,832 degrees Fahrenheit (6,000 degrees Celsius), the same temperature as the surface of the sun, and hot enough to generate an endless amount of electricity. The potential for geothermal energy is thrilling, yet the innovation gap has kept its fulfillment at bay. That is, until now. Over the past two decades, new technologies in the oil and gas industry, such as horizontal drilling, multi-stage fracturing, and managed pressure drilling, have led to exciting research and development breakthroughs in geothermal.

Entrepreneurs and startups are beginning to catch on to the potential for this technology transfer from the fossil fuel industry to geothermal, specifically in the state of Texas, where the second-highest policy concern among voters in 2022 was power and electric grid issues. In terms of politics, what gives geothermal unprecedented potential for success is its uniquely bipartisan support. Studies by the Union of Concerned Scientists and MIT indicate that geothermal is poised for the same kind of overnight, exponential growth that fossil fuels had in the Industrial Age. If that's not a thrilling prospect, then I don't know what is.

Although geothermal energy has for the most part managed to stay a politically neutral topic in the climate change conversation so far, it's most likely a matter of time before political divisiveness will ask people to pick sides. As young innovators, thinkers, and voters, it's up to us to learn more about new technologies like geothermal and in turn help educate others on their benefits and drawbacks.

Many of geothermal energy's potential issues are already being resolved thanks to continued research and development. For example, deep drilling for geothermal energy has been shown to cause small earthquakes as well as the risk of pollution and altering the landscape through fracking. Luckily, these issues are already being resolved as scientists find methods that require shallower drilling and lower temperatures to generate the same results. At the time of this writing, geothermal is being produced in only a few parts of the world, including Iceland, Italy, Japan, New Zealand, Russia, and the US. It accounts for only 0.1 percent of the world's energy production. However, as I've already said, that number will most likely grow very soon.

Now that you've seen for yourself why the current status-quo plan for climate action isn't working, let's take a deeper dive into The Climate Commitment, a plan that will. As you'll see in the following chapters, young people are poised to lead the way in a clear-cut direction that embraces local, free-market, and nature-based solutions that conservatives, whose voices have long been missing in the climate space, can embrace.

The 411 from Chapter 3

- We rely far too much on China for renewables production and lithium battery mining and manufacturing, as well as on Russia for liquid natural gas (LNG).

- Continuing to rely on our adversaries for energy and manufacturing has a definitive negative impact on our environment.

- People want green energy until it bothers the landscapes in their backyard. NIMBYism is the great hypocrisy of American climate action. We need to compromise and embrace solutions that work.

- The US ranks number one in installed nuclear energy capacity, positioning us as clean energy leaders, and we cannot lose that position.

4

Protecting Local Communities for Global Results

n 2020, I set out in a Tesla X for a grueling forty-five days across thirty-six states to highlight (through thirty-five podcast episodes) what small businesses and public-private partnerships are doing to combat climate change. We called our project the *Electric Election Roadtrip* and on it covered the most rural inroads of our country, the coastal cities, and every corner in between. I saw for myself how for generations, innovation and market-driven solutions have been working to improve our environment at the local and state levels. I also witnessed how, considering the varying geography and economic factors of different communities and their diverse local government infrastructures, one-size-fits-all solutions simply won't work.

Take, for example, composting. For a middle-class family living in Massachusetts whose town has provided a composting bin for their backyard and has designated free weekly pickups to take their compost to a nearby processing facility, it is relatively easy to do their part. But a farmer living in a West Virginia county with no composting program has no knowledge of where or how to begin, nor is there any infrastructure in place to support him. Until the local government steps in with educational programs, available equipment and supplies, and financial resources, any composting initiatives at the federal government level will be in vain.

Even among communities that have well-established composting programs and facilities, the variables span a wide range in terms of impact. In Seattle, where I lived during and after college, for instance, our state-of-the-art processing plants can handle even wastes such as plastics and meat bones, whereas another county in Washington State may have a much more limited capacity. (The way to gain more funding and support in these communities, by the way, is to ask the government at the local and state levels first. Only when those resources are exhausted does it make sense to take it to a federal representative who can advocate for a particular community.)

It's impossible for a five-hundred-page federal government bill to handle all of the intricacies of our vastly different communities and their needs. So why are we always starting at the top to try to solve climate change when we should be starting at an individual and local level first? Why not tap into the rich resources of the private sector and the entrepreneurial spirit that has characterized America as a world leader of innovation?

Even though the Right has always held this message as a core value, for too long we have shied away from applying it to the issues surrounding climate change.

Meeting the World Where It's At

We live in a vast country with an array of climates, landscapes, and biomes. Climate action that works for someone in one region doesn't necessarily work for someone in another. The top-down regulations the GND proposes have never addressed what is reasonable (or even possible) at a local level. Based on a number of factors specific to each region, what might be possible in one locale may be inconceivable in another.

Seattle, where I lived for six years, is a coastal city with such abundant access to hydropower that getting off fossil fuels would be a relative breeze there compared to trying to do so in a landlocked city in, say, the Midwest, where I grew up. Similarly, attempts to ban the use of plastic bags might seem feasible in high-income areas, but it is far more difficult for a farming community in Appalachia, where a five-cents-per-paper-bag surcharge is an added burden to many families living paycheck to paycheck.

We don't need to look further than our own backyards for proof that a one-size-fits-all approach is not the answer. Democratic governor Gavin Newsom's plan to make California a carbon-neutral state by 2045 by relying on solar, wind, and hydropower has turned into nothing short of a disaster. Despite its bold posturing as a world climate action leader, the state has repeatedly been forced to institute rolling blackouts—planned power outages in response to an electricity demand that exceeds the supply capability—and the governor now finally admits that the state does not have enough energy sources on hand to sustain an emergency. To make matters worse, California must allocate hundreds of millions of dollars to buy power from the same fossil fuel plants that are scheduled to shutter. The state has also had to borrow energy from power plants in Arizona and Utah, such as the San Rafael coal facility I will expand on shortly, and has finally been considering reopening a nuclear power plant that has been closed for several years.

If we understand climate change as the result of countless trade-offs in the name of advancement, it's easy to see that finding a solution means deciding on a new set of trade-offs that will restore balance to our planet. Fossil fuels have led to great progress. But they have also taken a toll on the environment via carbon emissions and other pollution-related issues. Put another way, making stuff and going places faster and cheaper has greatly benefited and also greatly damaged our planet. The only way to fix this damage is to make new choices that will, similarly, have their own set of risks and rewards.

Considering how convenient, abundant, and affordable fossil fuels are as an energy source, it's understandable why about 80 percent of the world's energy supply comes from them. Since their widespread adoption, human health and welfare have improved remarkably, poverty has been greatly reduced, and our population has increased from one billion to eight billion in just over a century.

And I'm not only talking about energy. The petroleum that's derived from fossil fuels is found in virtually everything we use on a daily basis. It's in all plastic and rubber products—cups, phone cases, eyeglasses, new construction, food packaging, shipping materials—cosmetics, hair dye, shaving cream, soap, chewing gum, aspirin, and 60 percent of clothing worldwide. The food industry uses it as a coating on certain cheeses, raw fruits and vegetables, and chicken nuggets. Even your grandparents' dentures are made from acrylic resin rendered from—you guessed it— fossil fuels. We literally eat, sleep, breathe, and chew with fossil fuels.

And finally, here's one that might surprise you the most. Did you know that solar panels, the very same technological advancement that is being touted as the greatest alternative to fossil fuels, are actually made from fossil fuels? Eleven tons of coal must be burned for the energy used to make a single solar panel, and every step in its production and deployment requires an input of fossil fuels—from the carbon reductants needed for smelting silicon to the intercontinental transport of materials, and on-site installation.

This brings us back to the discussion of trade-offs versus one-size-fits-all solutions. Intuitively, most of us know that risks and rewards are embedded in every choice we make about the environment. Yet politicians and talking heads present a much more black-and-white picture when it comes to climate change, a picture that our generation needs to take the lead in correcting once and for all. In an interview with Representative Dan Crenshaw (R., TX), who has long been a champion of Texas's environmental initiatives, the congressman insisted that this extremist approach has to change. "Rational discussions about trade-offs is a governing official's job fundamentally. And Americans should start demanding that from their elected officials."

To demonstrate how important it is to include the concept of trade-offs in finding local climate solutions, let's consider the Texas energy sector. Texas leads the nation in energy production, accounting for about 12 percent of our total net energy generation. In 2021 the state was the largest producer of oil (43 percent), natural gas (25 percent), and wind-powered electricity (26 percent) in the nation. Its fossil fuel industry employs about 14 percent of Texan workers, according to the American Petroleum Institute.

So what would happen if Texas stopped producing fossil fuels? The answer points to a stark reality. We would be forced to import a much dirtier version of coal, oil, and gas from other countries, such as Venezuela, Russia, and China. Not only would we be raising emissions levels if we did this, but we would also be benefiting the economies of these other countries instead of our own. And, if we consider how much of the workforce of states such as Texas is tied to the fossil fuels industry, our country would face a huge unemployment crisis to boot.

For these reasons alone, banning fossil fuels country-wide just won't work. Yet, the GND assumes that wind and solar power will soon do an adequate job replacing fossil fuels everywhere. Forgive me for stating the obvious, but wind and solar only work when the wind blows and the sun shines. In many parts of the world and for many months of the year, both of these elements are not readily available. Not to mention that, as we have just seen, the production of wind and solar systems is totally reliant on fossil fuels.

A look at outcomes in other countries that have already tried to replace coal, oil, and natural gas with only one or two alternative energy sources proves the inefficacy of this strategy. Take, for example, Germany's *Energiewende* (energy transition) that in 2011 called for the shuttering of its nuclear plants and the scaling of wind and solar, with the goal to reduce the country's carbon emissions by 80–95 percent by 2050. Similar to the GND, the plan was also meant to create hundreds of thousands of green-collar jobs via this transition. However, due to delays in building the proper infrastructure to store and transport renewable energy and unpredictable fluctuations in weather patterns, Germany has faced dire energy shortages.

On top of this, Germany has been forced to import natural gas from Russia and is now eagerly awaiting the completion of Nord Stream 2, a natural gas pipeline that will permanently link the two countries. Furthermore, this imported gas, although Russia bills it as *renewable*, incorporates wood burning into the process and ultimately emits more carbon than German coal, the use of which is also being revived to offset the demand for gas. Needless to say, Germany has failed to meet its decreased emissions goal.

We need a climate action plan based on fact, not fiction, one rooted in the understanding that every choice is a trade-off. We need to have realistic goals with initial impacts that may seem small, but over time will prove sustainable and universal. We need to stop the virtue signaling of entire geographic regions and acknowledge that global warming impacts the entire globe, and therefore its solutions must do so as well.

The unrealistic call to end fossil fuels altogether is only one of many examples of why we need to take a more holistic, localized approach to climate action. The problems are complex; therefore, so are the solutions. The sooner politicians start talking about solutions in terms of trade-offs, the sooner we can embrace a whole portfolio of options based on costs versus benefits.

According to Representative Dan Crenshaw, it's time for a renaissance of American innovation: "For a long time Republicans have felt they were on the defense on the topic of climate change. Instead of getting creative and getting on the offense." And who better to lead the way with this charge than our generation?

The Future of Coal

The necessity of protecting our local communities using innovative solutions is most clearly portrayed in discussions of coal. With at least 45 percent of all American coal set to disappear by the end of the decade, coal communities are among the hardest hit by the transition to clean energy. While a decrease in coal production is an important part of any plan to lower GHG emissions, the way policymakers have gone about it in the past has led to a disastrous impact on coal-reliant communities. In the last decade, mandates such as Obama's Clean Power Plan led to dramatic job losses and left communities economically devastated.

In New Mexico, for example, while Democratic leaders celebrated the recent shuttering of the San Juan power plant, the realities of what this means for the surrounding communities, especially the Navajo Nation, have started to set in. Hundreds of jobs have evaporated, along with tens of millions of dollars in annual tax revenue previously used to fund schools and community colleges in the area. Although the closure will reduce air and water pollution, including methane from the oilfields that has reportedly caused health problems for residents, debilitating economic loss has rippled through every facet of life, forcing many families to move away from their tribal community and search for work elsewhere.

How can we keep coal-reliant communities thriving during the transition to renewables? Many coal plants like the one in San Juan are being considered as sites for solar and battery storage projects. By repurposing these sites, builders can bypass the challenge of red tape in building new renewable energy infrastructure. Right now, the influx of proposals to build new wind and solar projects has slowed down the approval process to up to four years. If, however, companies can use existing transmission lines, the turn-around time for going online becomes a lot shorter. Additionally, communities that would normally object to new wind or solar farms on undeveloped land would put up less resistance to a reused plant that is already there. Yet another benefit of the plan to repurpose old plants is the existing electric infrastructure—the power lines, the transformers, and all the equipment that's needed to put electricity on the grid—that is already in place.

The transition to clean energy is already happening at dozens of other coal plants across the country. In Illinois alone, eleven plants will close over the next three years and be converted to solar farms or battery-storage facilities. In Louisiana, where a coal plant closed in 2022, a new solar farm is planned that could power forty-five thousand homes. In Hawaii, where the last-ever shipment of coal arrived in July, a huge battery storage facility is now being built with Tesla megapack batteries near a former coal plant. In Massachusetts, coal plants near the coast will soon connect with offshore wind power.

On a visit to the San Rafael Energy Research Center in Emory County, Utah, with Representative John Curtis (R., UT), I learned about the facility's transition to clean energy in a community whose economy has relied on coal production for a century. The Center is an excellent example of innovation and collaboration as it continues to provide a baseload of energy to supplement solar and wind power while at the same time researching and testing all kinds of other sources of power, such as wood and moss.

While the reuse of shuttered coal plants sounds promising, it is not a silver-bullet solution. Some of those projects have been delayed due to supply-chain problems, which I will discuss more in Chapter Six, and others are on hold indefinitely because of historic inflation and other economic constraints. Even without these delays, renewables aren't a one-for-one replacement for a coal power plant in terms of employment opportunity, energy production, economic growth, and other factors. Solar and battery storage projects require fewer workers; at the same time, they require more space. As I've already said, new solutions require new trade-offs.

And let's not forget nuclear. In a 2022 report sponsored by the DOE's Office of Nuclear Energy, 157 retired and 237 operating coal plant sites were investigated as potential candidates for a coal-to-nuclear transition. Evaluated parameters included population density, distance from seismic fault lines, flooding potential, and nearby wetlands, to determine if they could safely host a nuclear power plant. The study found that 80 percent of the potential sites were indeed suitable for hosting advanced nuclear power plants.

At the regional level, replacing a large coal plant site with a nuclear power plant of equivalent size could provide some 650 jobs across the plants and supporting supply chains and $275 million of economic activity. Most nuclear jobs typically come with wages that are about 25 percent higher than any other energy technology. In addition, reusing existing coal infrastructure for new advanced nuclear reactors can lead to average construction cost savings of 25 percent.

Specifically, small modular reactors (SMRs) that are now being built privately by non-government entities are uniquely positioned to redirect skilled workers from the coal power plant industry to new nuclear power plants. A single SMR, for example, could provide at least 237 on-site jobs, which is much more than the capacity of a typical coal plant. These jobs could last throughout a forty-plus-year life of the plant with a median wage scale roughly 60 percent higher than a job in renewables. In addition, roughly 1,600 additional jobs could be created for an estimated three-year period of construction of an SMR.

In Chapter Seven I will discuss the benefits of nuclear and

other clean energy possibilities in greater detail. The important point is that the coal, oil, and gas industries of yesterday's America do not have to be perceived as the enemy. Just as we've seen that true progress at a national and global level happens when unlikely forces join to solve a problem, so too can collaboration between the old and new guard in the energy sector bring unexpected solutions as we evolve toward a green future.

What Real Progress Looks Like: Engaging Fossil Fuel Communities

Of the countless smart young people I've spoken to who deeply care about climate change, very few can articulate any solutions beyond the common sound bites on their Twitter or TikTok feeds: *Ban fossil fuels! Switch to wind and solar. Go, electric vehicles!* If I probe much further about how to make the proposed giant leap to clean energy without disrupting giant swaths of the American population, or where we can ethically source the minerals to build enough wind and solar panels, or how we can afford to manufacture batteries for enough EVs to do away with gas cars, the conversation usually comes to a screeching halt.

Unfortunately, our generation's tutors on these issues are climate change celebrities who implore us to care more, give more, and do more. At the same time, we should consume less, use less, and waste less. Leonardo DiCaprio has become a spokesperson for several environmental organizations, including the World Wildlife Fund and his own self-named foundation, dedicated to raising awareness about the climate crisis. The actor also starred in

Don't Look Up, a Netflix satire that pokes enormous fun at divisiveness on climate issues. In the movie, a fast-approaching comet is the chosen metaphor for the imminent danger of global warming and the need to take immediate action against it. The actor touts the film as "a unique gift to the climate change fight," a self-aggrandizing statement and a head-scratcher at once. If you look for any policies or practices that are a direct result of Mr. DiCaprio's work in environmentalism, I guarantee, you won't find much.

Virtue signaling is bad enough. Add to it blatant hypocrisy, and it will get any thinking person riled up. Billionaire climate showmen who call for drastic lifestyle changes, including renouncing fossil fuels altogether, can't help but invoke outrage as they soar through the air in private jets that put out over 1,760 tons of CO_2 into the atmosphere per year. Despite their generous annual donations to supposedly offset their carbon footprints, such extravagant environmentalists can hardly be taken seriously by Americans who struggle to make car payments and pay mortgages—and are told we will soon have to switch to unaffordable energy options to heat our homes or get to work.

The reality is that most nations, including ours, rely on fossil fuels to power our homes and businesses. Therefore, we need to focus first on decarbonizing fossil fuels rather than abruptly banning them. Renewables such as solar and wind are good sources of intermittent energy when the sun is shining and the wind is blowing. Still, our reliance on China for mining and panel and turbine production continues to be problematic in terms of ethical, environmental, and national security standards.

Cleaner, more reliable, and scalable energy sources exist, such as nuclear, hydropower, hydrogen fuel, and geothermal, and it's only a matter of time before they become more affordable. By streamlining the permitting process, reducing unnecessary regulation, and instituting powerful incentives that encourage innovation, we can speed up the development of these solutions. As we will continue to explore throughout the book, American entrepreneurship is poised to make incredible advances in climate tech and the clean energy sector. The fastest way to tap into this potential is to free it from the bureaucratic red tape that has been holding it back.

It's a fact that countries leading in emissions reductions have some of the most free-flowing economic markets in the world. From large-scale think tanks to coal mines that have converted to clean energy production to small businesses that are finding effective ways to reduce emissions and enhance carbon capture, innovation is taking place every day across our nation. Beginning locally and through public and private partnership, we can unlock resources and incentives to scale this work and encourage more research and development for other energy sources.

What if our new race to the moon were the race toward American climate solutions that stimulated job growth within fossil fuel communities? What if we could see climate change as a market opportunity that would serve both the environment and the economy at the same time? Because the global annual energy market, which at about $7 trillion is roughly a third of the US GDP, is already moving toward clean energy, wouldn't it benefit our country to lead in that movement, instead of continuing to concede that role to China and Russia? Collin O'Mara, the

President and CEO of the National Wildlife Federation, puts it in the simplest terms I've heard yet: "If climate action equals jobs, we win!" More green jobs for Americans would certainly be a measurable, targeted, balanced, adaptable solution supported by real people.

In 2021, the artists behind the Climate Clock decided to infuse their display with what they considered a note of optimism, by adding a second set of numbers that represents the increasing percentage of the world's energy that comes from clean sources such as sun and wind. They also added this message: "The Earth has a deadline. Let's make it a lifeline." Though the addition infuses the message with a more positive tone, it's not altogether accurate. While it's true that more and more countries across the globe have been adopting cleaner energy sources, this bit of progress is only one part of the equation.

The good news is that progress is happening in ways that would shock most people. For example, the United States has decreased its carbon-related pollution more than any other country since 2000. Four-fifths of our states have managed to decouple their CO_2 emissions from economic growth, showing us that scientifically sound climate action doesn't have to cost us trillions of dollars.

Nor do solutions have to be sexy to be powerful. No one knows this better than Curt Eliason, the Vice President of RENEW Energy Maintenance, LLC, who has spent the past twenty-three years repairing wind turbine components in South Dakota. Just like my grandmother, and many other Midwesterners, Curt has always been a conservationist at heart, though he never formerly called himself an environmentalist. Now that he is involved in the renewable energy business and is raising Gen-Z kids, he calls himself both.

"Why throw something away if we can use it for another purpose?" Curt asks. "A lot of South Dakotans share that same mindset. We are not wasteful." That goes for wind as much as anything else. Having grown up on a ranch in Sioux Falls, he has been inherently aware of the rich wind resource around him, which he believes is enough to one day power the whole country. So, while fixing gear boxes when they break down might not be on every young person's top five list of career options, when you consider the US's sprint toward wind power, it is more than likely a smart choice. With states such as Texas, California, and Indiana using RENEW's services, it is clear that the demand for expertise in wind turbine engineering is high across the country.

But an abundance of clean energy is nothing new for this part of America. RENEW is only part of a decades-long legacy of climate-friendly businesses and projects spearheaded by South Dakotans, including the installation of hydroelectric dams in the Missouri River in the 1950s. South Dakota's clean energy revolution has placed it third highest in the nation for the number of wind-powered homes. What's more, wind has surpassed hydroelectric power as the largest generator of electricity in the state, with over 52 percent of electricity generated coming from wind turbines in 2021. South Dakota now produces twice as much electricity as it uses and exports the rest of its clean energy to nearby states.

Even though wind produces the majority of electricity for

South Dakotans each year, it doesn't do so every day. In fact, that day-to-day number of units varies greatly because engineers still haven't developed efficient ways to store wind energy for usage when the wind isn't blowing. This fluctuation in available wind power makes reliable backup sources such as nuclear, hydropower, and natural gas imperative. (More on this later!)

Even for a state with a robust clean energy output, it's unrealistic to think we can completely do away with oil and gas. Unfortunately, in the name of limiting emissions, many of our production and manufacturing industries have been forced to shut down their facilities and resort to offshoring fossil fuels. This type of practice makes no sense economically or environmentally. For many rural communities, that single steel mill or power plant was the main source of job security for many generations of families—and now it's gone. In addition, our continuing reliance on commodities such as fossil fuels has forced us to buy dirtier sources of energy in other parts of the world. Alienating the fossil fuel industry and dwelling on the sector's negative impact is only hindering the climate progress we need to make. Instead, we should highlight the fossil fuel industry's commitment to reducing emissions over recent decades, making it already up to 40 percent cleaner than that of Russia and China. In October 2022, along with two other members of the ACC team, I decided to visit Midland, Texas, where much of our country's oil comes from, as does President George W. Bush. As we toured first the downtown area and then some of the city's smaller to middle-sized oil and gas companies, I couldn't help but notice that everything, including the local bank's logo of a drill, spoke of oil and gas. These people were proud of their oil-and-gas legacy, not because they despise the planet, as many radical activists would have us believe. They were proud because of generations of labor their families have devoted to powering America's homes and businesses.

As some Midland residents expressed to me, oil and gas workers feel attacked and vilified by the same people their industry has helped usher toward prosperity. They acknowledge the need for cleaner energy and would like to be a voice at the table; instead, they feel completely shut out of the climate policy conversation. It's no wonder that we could feel the community's visceral prickliness at having environmentalists on their turf. I knew enough to recognize that this response was nothing personal, but rather a measure of how far leadership still has to go to build a spirit of collaboration with the so-called "enemy."

One way we can continue to bridge the gap between climate action and the fossil fuel industry is through creative incentives such as monetizing carbon, an idea economists and private companies have been grappling with for years. Nori is a unique startup that has managed to do so, combining climate action with cryptocurrency to create the world's first blockchain-based marketplace for carbon removal. Buyers are encouraged to first calculate their individual or business's carbon footprint to decide how much to invest. One NRT (Nori Carbon Removal Tonne) token purchased represents one metric ton of carbon dioxide removed and stored for a minimum of ten years.

Carbon removal is not the only hot ticket in incentivized

climate action. Companies are also capitalizing on smart ways to turn trash into treasure. In the United States today we generate an excess of 290 million tons of solid waste annually, about 40 percent of which is estimated by the EPA to be organic and waste paper. BurCell Technologies has developed a system to divert this waste away from landfills by recycling it into an energy-rich feedstock that turns out to be more digestible than other feedstocks. This win-win case is one of many examples of entrepreneurs sidestepping the alarmism versus denialism trap to achieve great improvements.

What Real Progress Looks Like: Engaging Agricultural Communities (Including "Big Ag")

Activists frequently decry the industrialization of American farming, derisively referred to as "Big Ag." In 1800, more than 70 percent of the US labor force worked in agriculture, and as late as 1900, some 40 percent still worked on farms. Compare these numbers to today's figure of less than 2 percent and you can easily see the effects that consolidation and technological advancement have had on the industry.

By investing in labor-saving and productivity-enhancing practices and technologies, farmers have been able to work larger plots of land and produce larger harvests—enough to feed the rest of the population—with fewer people. (Still, agriculture accounts for nearly 17 percent of employment in highly rural and remote areas of the United States.) This type of progress is the hallmark of all thriving societies. In fact, no nation has ever succeeded in moving most of its population out of poverty without most of that population leaving agriculture work.

Large-scale US farms, because they are so economically and environmentally efficient, are also able to produce vastly more food than Americans can consume, making the US the world's largest agricultural exporter. That benefits the US economy, of course, but it also comes with an environmental benefit for the world. A pound of grain or beef exported from the United States almost always displaces a pound that would have been produced elsewhere with more land and GHG emissions.

Therefore, Big Ag is not necessarily the enemy. Neither is conventional farming. Contrary to what many people might think, organic farms, both large and small, don't actually outperform large conventional farms by many important environmental measures. Because organic farming requires more land for every pound produced, a large-scale shift would entail converting more forest and other land to farming, resulting in greater habitat loss and more GHG emissions.

The EPA estimates that agriculture accounted for 11.2 percent of US GHG emissions in 2020. But the benefits of research and development in fertilizers, crop breeding, farm equipment, and land usage can reduce emissions both from farm inputs like fertilizers as well as from land-use change. Advances in crop breeding, farm equipment, and other agricultural areas in recent years have helped farmers grow more food with less fertilizer and land, and thus less land-use conversion and GHG emissions. Because of such innovations, the carbon intensity of agriculture fell more than

10 percent between 1990 and 2015. And yet, the GND allocates only $601 billion in helping farmers transition to greener practices. That's only 3.5 percent of the total deal.

Why don't more farmers embrace green practices? Simply because many don't know about them. Nor do they know that many bills exist that actually provide government incentives to farmers for adopting such practices. Instead, most farmers get their information from the giant agricultural businesses that have taken over most of the world's farming. These companies have so much economic control over farmers that even if a farmer wants to switch to more environmentally friendly practices or participate in a government-incentivized program, many times, they feel it is too risky. Until "Big Ag" companies start to embrace some of these innovations themselves, many farmers won't be able to either.

There are many more climate-friendly farming practices that the government can and should incentivize that would yield excellent environmental benefits. Planting cover crops, for example, provides continued feeding and stabilizing soil before and after the main cropping period of large crops such as soybeans and corn. These cover crop plants are not meant to be harvested, but rather are intended to manage the soil, protecting it from erosion, weeds, pests, and diseases, and promoting biodiversity, wildlife, and soil fertility and quality. They also provide continual carbon sequestration throughout the year, not just in harvest season. Cover crops could also provide good bee foraging material, which would benefit the honey industry as well.

Taking into account the competing forces that influence the

global farm and food system and how these drivers influence long-run agricultural land use, production, prices, GHG emissions, and food consumption, research has found that doubling agricultural research and development between 2020 and 2030 would reduce global agricultural land use by about sixty-three thousand square miles. That's about the size of Iowa. It would also reduce the level of emissions from fertilizer and fuel use per unit of food produced by about twelve percentage points. In effect, GHG emissions would be reduced by over one hundred million tons—equivalent to one-sixth of current US agriculture emissions—while also lowering global food prices by about eight percentage points.

Because the US is a major global food producer, these benefits will naturally spill over to urban communities as well as the rest of the world. Productivity gains give the US a larger share of global food production by increasing farmers' international competitiveness. This gain reduces global environmental impacts because US farming is typically more resource-efficient than farming elsewhere. Ultimately, by increasing public agricultural research and development funding in this country, we would help address climate change, land-use conversion, and habitat loss, as well as food insecurity.

Timber production is another rural and agriculture-based industry that is finding ways to help solve climate challenges while still growing their business. Vaagen Timbers in Colville, Washington, is blazing a new trail in timber production with their cross-laminated wood panels made from the by-products of forest thinning and restoration. As you will learn more about later, tree thinning is a powerful practice that promotes healthy forest growth by removing defective or slow-growing trees to make room for strong trees. It is also an important tool in forest-fire management that needs to become a top priority if leaders are serious about reducing carbon emissions. (Much more on this later.)

The beauty of Vaagen Timber lies not just in their zero-waste production that leaves behind healthier forests, but also in the strong coalitions the company has formed with national conservation groups, forest services, local businesses, and Congress. By using science and technology to mimic what nature already does on its own, they are replacing carbon-intensive steel and concrete construction of our buildings.

Examples of this type of innovative technology abound in rural communities. Across the country, many other family-owned businesses like Vaagen Timber have found ways to join the fight against climate change and prosper economically at the same time. This type of collaborative effort between the private and public sectors is where much of our progress against emissions still has to be made. It's vital that we continue to share these stories and raise awareness about what's possible.

WHAT IF . . .?

Innovation begins with asking questions. What if carbon were considered a commodity instead of a waste product? Companies such as Texas-based NET Power that are on the cutting edge of carbon-capture technology are making this concept a reality. The science of recycling carbon is complex, but to keep things simple, the method NET Power employs goes something like this: Natural gas is burned with pure oxygen and results in CO_2. In turn, this CO_2 undergoes a multi-step recycling process that involves a combustor, turbine, heat exchanger, and compressor. This process is surprisingly cheaper than traditional technologies while also producing zero emissions. The captured CO_2 is also pipeline-ready and can be either inexpensively sequestered or, better yet, sold to industries such as the medical, agricultural, and industrial sectors.

Progress is being made in the private sector with many out-of-the-box ideas. We need to allow businesses to continue to innovate and then support projects when they are ready to scale at a state and national level. How can the government help stimulate this innovation? By incentivizing it with more provisions like 45Q, which provides a tax credit for carbon sequestration. Innovation happens from the ground up, not the other way around.

As we covered earlier, the Green New Deal fails to capture the complexities of the issues at hand, including the importance of sustainable agriculture and other natural climate solutions as part of the solution-set. Instead of *revolutionary* policies, we need an *evolutionary* approach to climate action, one that will lead to new solutions that account for the complexities and needs of our communities.

The following are just a handful of alternatives to fossil fuels that need continued support and research:

- The development of American critical minerals and the building of solar panels in America instead of China, Russia, or the Democratic Republic of the Congo
- Producing oil and gas responsibly. This can be done by providing regulatory certainty, maintaining fair leasing auctions on federal lands, and encouraging private sector investment
- Increasing the extraction of cleaner American natural gas to power our economy and ensure energy security at home and abroad
- Securing nuclear fuel supply chains by mining more uranium domestically and working with our allies instead of Russia
- Expanding hydropower and geothermal energy through greater investment and clearer regulations

These are the concepts we need to be discussing with our communities and supporting at local and state levels.

An evolutionary approach to climate action means out-of-the-box thinking that works at a local level and for a specific geographic location. The Block Island Wind Farm, the site of America's first five offshore turbines, is the perfect example of this type of custom-made innovation. After a local vote in favor of the project, and thanks to the collaboration of clean energy technicians, engineers, and over three hundred local laborers and the support of Rhode Island state politicians, the construction of the wind farm took only two years to complete from start to finish.

It is up to each state to decide which trade-offs are worthwhile and which ones aren't. For a landlocked state such as Arizona that is faced with the challenges of dry heat, wind is a lot less viable than energy from, say, solar, nuclear, or natural gas. Until recently, Arizona was also using hydropower generated by the Hoover and Lake Powell Dams, yet because of rising temperatures, those sources have significantly dried up, and now the state is struggling to find new sustainable energy sources that will continue to move them away from coal. Needless to say, Arizona's trade-off conversation will look a lot different from that of Rhode Island.

The GND, with its platitudes, gross generalizations, and unsubstantiated promises, skirts the all-important conversation about trade-offs that needs to happen for progress to be made. We need to focus on the smaller, local victories first before we can scale to sustainable global solutions. In a memorable *Electric Election Roadtrip (EER)* interview with Mike Carloss, the director of Conservation Programs at Ducks Unlimited (DU), situated along the Mississippi Delta in Louisiana, I saw firsthand how starting small can have a big impact. For decades, DU has worked hard to restore and enhance the precious water channels that are home to thousands of species of birds, fish, and other wildlife.

As Mike puts it, "We do what's right for the ducks." In this case, what's right for the ducks is also what's right for the communities that share their wetlands home and whose livelihoods have been teetering on the disappearing coastline. By working with local volunteers, businesses, and gamesmen who care equally about preserving such a rich natural resource, DU has managed to protect and grow over fifteen million acres of valuable wetlands.

Who else is one of the DU's biggest supporters, but Cono-coPhillips? One of the largest American oil and gas companies, ConocoPhillips owns 643,000 acres of coastal wetlands in their region and has donated generously to their programs and lent valuable infrastructure resources to support the group's efforts. If this type of unlikely partnership in climate action isn't out-of-the-box thinking, then I don't know what is.

Let's Give a Crap about Our Future

There is another reason why starting on a smaller scale works better than a top-down approach. There is simply less red tape. According to Martin Durbin, president of the Global Energy Institute at the US Chamber of Commerce in DC, our permitting system at both a federal and state level often cripples our ability to make meaningful climate action. In an *EER* podcast interview, Martin explained, "If you want to build a highway, it will take seven years to get just a *yes* or *no* answer . . . We need to be able to build big things quickly. Speed is the common denominator."

Local communities working together with private sectors simply have more power to move quickly in innovation and produce more immediate results. Brightmark, a bold new startup company, one of 132 similar ones in the state of Wisconsin, is a shining example of this type of partnership. The company has found a way to help local dairy farms not only produce zero waste, but also reduce emissions by turning cow poop into a clean, sustainable biogas for local farms. At the same time, the company creates a baseline income that helps farmers weather the peaks and valleys of market price fluctuations. The by-product of cleaned-up fiber can go back to these dairy farms to be used as bedding or manure, or it can get bagged and sold for extra revenue. On top of this, the low-phosphate by-product from this process actually reduces algae bloom and cleans up our waterways as it passes into the ocean.

Speaking of oceans and poop, the excrement from whales is yet another unlikely resource scientists are looking into to help significantly lower Earth's harmful gases. About a decade ago, a German scientist discovered that whale poop releases nutrients at the surface of our oceans on which tiny living creatures called phytoplankton feed. Why is this important? These phytoplankton absorb nearly 40 percent of the world's carbon dioxide (equivalent to 1.7 trillion trees) and in turn release oxygen, in much the way trees do on land. In recent years, this naturally occurring carbon capture technology has caught the attention of other scientists, who are now experimenting with the idea of manufacturing artificial whale poop to encourage more phytoplankton growth, including the RPrime Foundation in Seattle.

This type of partnership between science and nature is the fabric of American innovation. In addition to upcycling cow and whale poop, hundreds of businesses are taking the lead with creative solutions such as man-made carbon capture technology, which, admittedly, is not as much fun to say as *whale poop*. Companies such as Oxy (short for Occidental Petroleum) are developing large-scale operations around the world that separate and utilize CO_2. With approximately 2,500 miles of CO_2 pipelines and over 6,000 injector wells, the company safely stores about twenty million tons of CO_2 each year. Oxy is also developing direct-air capture, with the plan to open the first phase of its one-million-ton facility in 2024.

As we have seen, solving climate change is about solving it from the bottom up, first locally, then nationally and globally, one challenge at a time. It's about having the right conversations about trade-offs. And finally, it's about applying a local, customized approach that supports American entrepreneurs and scientists. Next, we'll look at how to harness the free-market economy to create innovative, sustainable solutions to climate change.

The 411 from Chapter 4

- Local communities understand how to take care of their communities best. We need to empower localities to make decisions applicable to their own areas.

- State leaders (which surprisingly include Florida governor Ron DeSantis) are quietly making a big difference with bipartisan public-private partnerships.

- Compare that to states like California that have spent billions on EV mandates and have little progress to show for it.

- Banning fossil fuels is unrealistic. Look at the debacle of Germany's *Energiewende* as proof. Instead, we need localized solutions that work for individual environments, economies, and cultures.

- We need more research and development in cleaner American natural gas, hydropower, and geothermal, and mining domestic uranium, minerals for solar panels, lithium batteries, and more.

- We need to tailor clean energy solutions based on the resources and weather in different geographic regions (i.e., solar in Arizona, hydropower in Washington, wind in Iowa, nuclear in Georgia).

5

Utilizing America's Competitive Spirit to Build a Cleaner, Brighter Future

n September of 2019, when I was preparing to testify before Congress, I was still pretty green on the climate action scene (pun intended). I was twenty-one years old and, even though ACC was slowly taking off, I would have been the first to admit that I knew very little about running an organization. I knew even less about the scientific and statistical details of climate change. I assumed that America had a horrible track record of bad policy and uncurbed emissions because that's what I'd been told by the media and political leaders on the left. I was prepared to go into that hearing guns blazing and reprimand a group of leaders who seriously needed to be whipped into shape.

Luckily, just days before I was scheduled to speak, a team of experts briefed me on the most current environmental facts of our country, including where we stood compared to the rest of the world. As I perused the research material and heard the numbers, my jaw dropped. Contrary to my previous assumptions, the United States is in fact leading the rest of the world in emissions reduction, with a whopping 10 percent total decrease since 2005. Yes, we are responsible for about 14 percent of the planet's CO_2 emissions, which is a lot for one country, but it is still not as high as China's quickly increasing 27 percent.

It was a holy-sh*t moment for me—that soon turned into a double-header of a holy-sh*t moment. I was more than psyched to hear about the progress we'd made on the climate change front and proud of the concerted effort that had gotten us there. But my elation was quickly followed by a sobering thought: even if the United States went to a 100 percent decrease, with zero net emissions, the world's emissions would only go down by 14 percent, a mere drop in the global warming bucket. And even if our progress continued in leaps and bounds, the rest of the world's emissions most likely would still continue to climb without our help.

Since the start of the Industrial Revolution, economic output and greenhouse gas emissions have always risen in tandem. It's simple: a production increase equals a pollution increase. The good news is that, globally speaking, we have has finally broken the historic link between rising prosperity and CO_2 emissions. What this means is that the energy intensity of the world's GDP—the supply of energy needed to produce a dollar of national income—has fallen faster than the GDP has grown. For example, since 2005, even though America's GDP has risen by 29 percent, emissions have fallen by 15 percent.

Besides switching to cleaner energy sources at home, other

factors that have contributed to the decoupling of economic growth and GHG emissions in the West include lower emissions in countries where some of our energy has been outsourced. Emissions-trading schemes and different forms of carbon pricing have contributed to the curb as well. Technological advancement also plays an enormous role, allowing for the manufacturing of cleaner products at the same cost. For example, cars have gotten much more efficient since the gas-guzzling days of just a couple of decades ago. Growing up, my dad's Toyota Sequoia got a measly twelve miles per gallon, while that same model now yields an average of thirty. I'd say that's progress!

The fact that our country's greatest environmental strides in recent years have been achieved concurrently with America's greatest economic growth demonstrates that prosperity is not incompatible with effective environmental stewardship. In fact, I would argue that without the former, the latter is impossible. While it's true that the developed world is responsible for much of the damage we have done to the planet thus far, we are also in the best position to repair it. We often hear that we must decrease economic growth, population, or technology to stop climate change. This stands in exact opposition to the reality. Contrary to what radical activists think, market competition and technological development are our allies, not our enemies, in our fight against climate change. And every solution will require lots of market-driven capital.

In the last chapter, I talked about the untapped potential of our rural communities for economic stimulation and climate action. During one of my favorite *EER* podcasts, I was able to speak with Ron Allen, the Chairman and CEO of the Jamestown S'Klallam Tribe in Sequim, Washington, who has devoted the past forty years of his life to promoting growth opportunities for his tribe members while simultaneously restoring the decimated salmon population in the local Dungeness River. Remarkably, he has managed to do this by creating a vast network of partnerships with sport and commercial fisherman, local fish and wildlife organizations, farmers, timber workers, and local, state, and federal government officials on both the right and the left. Negotiating everything from industry practices to irrigation systems to the construction of culverts and the management of water temperatures, Ron has done unprecedented work across the public and private sectors, as well as across partisan lines.

Ron and his tribe did not wait for some large-scale government initiative to swoop in with a top-down approach to environmental action. Instead, in the truly Native tribal spirit of self-reliance and love of the land, his community did whatever they could to protect their own backyard, leveraging the resources and talent that were at their disposal. Only when the tribe was ready to purchase land to secure better revenue sources did he look to the government to secure a small grant. Today the tribe is the second-largest employer on the peninsula.

When I asked Ron to explain the key factors in his success, he cited mutual respect among the diverse stakeholders in the project. He also emphasized a shared interest in protecting the multibillion-dollar set of salmon-based industries that were in danger of disappearing without swift action. Similarly, if we are to make large-scale progress on climate change, then we cannot ignore the economic progress that must go hand in hand with it to make it sustainable. Again, this starts with individual and local effort.

Growing the Green Economy

For young people entering the workforce, there's no better time to be an entrepreneur. America needs our generation to fill the information and technology gap we are facing across the public and private sectors. State and local government offices need young, educated men and women to step up so they can be called on for sustainability planning and guidance. As it stands now, many of these positions are being filled by untrained staff members who are playing catch-up trying to stay up-to-date on the latest knowledge and trends surrounding climate change.

In the private sector, companies of all sizes need more employees who are trained to look through the climate change lens as well. As the world becomes more aggressive in its efforts to fight climate change, many economic growth opportunities are beginning to emerge specifically around climate action, leading to new jobs and competitive opportunities across all industries.

In the investment space, capital deployment in green businesses is growing in both sectors too. In 2021, for example, climate tech start-ups targeting sectors that are responsible for 85 percent of GHG emissions attracted 39 percent of investment worldwide. That number jumped to 52 percent in 2022. And what exactly is climate tech? Climate tech includes any technologies explicitly focused on reducing GHG emissions or addressing the impacts of global warming.

In the past few years, the federal government has helped fund new ideas in the green energy space. The bipartisan Energy Act of 2020, signed by President Trump, refocused the US Department of Energy (DOE)'s research and development programs on scaling up clean energy technologies such as advanced nuclear, long-duration energy storage, carbon capture, and enhanced geothermal energy. It also aimed at reducing emissions from both coal and natural gas power plants, industrial processes, and direct air capture.

In 2021, President Biden's Bipartisan Infrastructure Law, signed by President Biden, was the first infrastructure law in US history focused on addressing climate change. Its implementation advances a wide variety of infrastructure investments aimed at reducing emissions in America's transportation network to help foster American manufacturing of green technologies. Among the many efforts made possible by the Bipartisan Infrastructure Law, the United States Department of Agriculture (USDA) has begun work on a strategy to address the recent uptick in wildfires.

The USDA is also putting resources from the Bipartisan Infrastructure Law to work in closing the digital divide in rural America that I discussed in Chapter One. By making high-speed internet accessible to more communities, more farmers and small business owners will have access to the real-time information and new technologies they need to maintain a competitive edge.

While not perfect, the 2022 Inflation Reduction Act, which included a $369 billion investment in climate and energy policies, marked the largest investment in US history to fight climate change. The act provides countless grants, direct subsidies, and tax breaks for projects and businesses in clean energy, manufacturing, and innovation. This type of government support has given private investors more confidence in climate tech–related businesses, which means that venture capital dollars will flow more freely now into the green energy space. The IRA also includes better standards and labeling for product declarations and carbon impact.

Besides incumbent climate tech companies, thousands of new startups are already benefiting from government incentives. In 2020, I visited Greentown Labs, a forty-thousand-square-foot startup incubator for climate tech businesses, located in Somerville, Massachusetts, where physical products can be prototyped, tested, and then eventually scaled.

On my visit, I was impressed not only with its facilities but also with the collaborative efforts taking place there among designers, engineers, investors, politicians, and corporate partners. For example, Shell, the energy and petrochemical company, has run two Greentown accelerator programs and has served as a founding partner of Greentown's headquarters and its second location in Houston. I sat down with Dr. Emily Reichert, the CEO of Greentown Labs, to talk about partnerships with fossil fuel companies to win the fight against climate change. "It's a gigaton-sized problem that needs gigaton-sized solutions. These companies know how to do it. There's no point in demonizing anyone. Instead, it's all hands on deck to solve the problem."

One of the problems being solved at Greentown Labs is how

to cheaply store large amounts of solar and wind energy to power grids when the sun isn't shining and the wind isn't blowing. The six-year-old startup Form Energy Inc. has built an inexpensive battery that can discharge power for days using iron, one of the most common elements on Earth. The company hopes that if all continues to go according to plan, its one-hundred-hour iron-air batteries will be capable of affordable, long-duration power storage by 2025.

The intermittency of solar and wind power is only one of thousands of climate change–related issues that young entrepreneurs, scientists, engineers, and innovators across the country are addressing. At the 2019 EarthX conference in Dallas, ACC hosted its own event: *30 Under 30: The Green Generation*, the first large-scale event in the country to highlight young environmental advocates, business leaders, and innovators in the climate tech space.

Among the participants at the event was Freight Farms, a company with the mission to empower anyone to grow food anywhere through its line of modular container farms and farm automation software. Thanks to the company's creative AgTech solutions, communities across thirty-nine countries so far can grow fresh, healthy food year-round as part of the largest connected network of farms in the world.

Another nominee at the event was 4ocean, a nonprofit ocean cleanup group that hires full-time captains and crews to recover plastic and other man-made trash from the world's oceans, rivers, and coastlines seven days a week. For every bracelet purchased, the organization promises to remove one pound of trash. Air Company, another participant in our event, has found ways to turn CO_2 into a never-ending resource for product development. Manufacturing everything from perfume to jet fuel with the unlikely help of the planet's most abundant pollutant, the company's carbon utilization technology is serving as the blueprint for profitable, industry-wide decarbonization.

Another inspiring initiative we learned about at the ACC event was a nonprofit called Blue Latitudes, an environmental consulting firm for offshore energy projects. Key examples of their work include repurposing offshore infrastructure as artificial reefs and analyzing fisheries activity around offshore infrastructure. The firm also offers a web-based platform that helps regulators, fishery managers, and energy industry stakeholders predict how the site-specific removal, reefing, or installation of an offshore energy structure will impact fisheries.

It's worth noting here just how important outdoor recreation industries such as fishing, hunting, hiking, and RV camping also are to our economy. In 2021 alone, our nation experienced a record $862 billion output from such activities, making up 1.9 percent of the GDP. This number surpasses other important industries such as mining, utilities, farming, ranching, and chemical products manufacturing. Outdoor recreation also generates millions of quality, high-paying jobs for people interested in protecting our natural world.

And for young people who might not have an interest in business or the sciences, you can still have a great impact working within the public sector, which is currently short-staffed by those trained in climate change. As someone who spends a lot of time on Capitol Hill, I've seen firsthand how the offices of senators and congresspeople have been forced to make do with untrained staffers because of this shortage.

Greenwashing: It's a Problem

While economic prosperity and environmental stewardship can and should go hand in hand, many businesses are taking advantage of the green consumer, offering misinformation about sustainability and biodegradability. This practice is known as *greenwashing*. With so many claims of environmental friendliness saturating the marketplace, even the savviest customers find it dizzying to try to determine what's real and what is just some marketing ploy.

Take, for example, the pledge of carbon offsets. To negate their own emissions, some companies donate money to programs such as tree-planting projects that theoretically offset environmental harm through carbon absorption. But what happens when wildfires and droughts destroy some of these forests? Many other factors on the production end come into play when trying to measure output and absorption balances. Constantly changing product lines create fluctuating amounts of emissions, as does ever-changing consumer demand.

Then there are companies donating to environmental causes without changing a thing about how they do business. This corporate version of moral licensing psychologically buys a company the right to keep doing the damage they are doing to the environment, because they believe they have already done their monetary good deed.

Perhaps the worst form of greenwashing occurs when companies allow their marketing teams to take over their so-called sustainability programs. These marketers publicly announce lofty environmental goals with no power whatsoever to follow through with them. In fact, I've heard from countless corporate marketing departments that college-age interns are usually in charge of writing a company's annual sustainability report. Companies know that their consumers are not planning on reading any of these reports to check if they've met their benchmarks, so why put any effort or time (even paying someone an hourly wage, which interns often don't even receive) into them?

Car manufacturers are some of the worst culprits of unrealistic pledges, with companies professing plans to transition 100 percent to EV production by 2030. Anyone who knows what's happening behind the car manufacturing scene knows that this would be a nearly impossible feat, considering what I'm about to discuss in the next section. Executives for these companies blatantly admit—at least behind closed doors—that this complete transition isn't possible, at least for now. Other businesses claim they will be 100 percent *clean energy, carbon negative, water positive, zero waste,* etc. But how can we as consumers check the validity of such claims? Furthermore, we all know that the wind doesn't blow and the sun doesn't shine 100 percent of the time. So, what type of energy is being used during those off times?

The reality is there is no standardized way of tracking progress or ensuring that companies are even meeting any of their lofty environmental goals. Green America, a nonprofit aimed at helping consumers navigate greenwashing, recommends looking for descriptions on a company's website that outline exactly how a product is *green*. Certifications granted by credible third parties such as B Corp and Fair Trade can give consumers a little more confidence in a product's green claims. Yet, like everything else, these certifications are not foolproof.

Environmental, social, and governance (ESG) was originally designed to be a useful metric to help investors assess the health and future profitability of a company, but has now become a huge source of political controversy. The *environmental* criteria of ESG consider how a company protects the environment through tools such as corporate policies addressing climate change. The *social* criteria examine how it manages relationships with employees, suppliers, customers, and the communities where the company operates. And, lastly, *governance* deals with a company's leadership, executive pay, audits, internal controls, and shareholder rights.

One of the main challenges with ESG scoring methodologies is they tend to focus on how well companies manage their internal processes, rather than the real-world impacts of their products and services. Take PepsiCo, for example, the maker of soft drinks and snacks all over the globe. The company tends to score favorably on ESG assessments, specifically in areas such as health and safety policies and climate targets; however, these are only measured internally before the products reach the consumer. Whether those standards hold up once inside our refrigerators or on our dinner tables is a separate question altogether.

Another problem with ESG is the large variation between the scores and ratings from different third-party ESG providers, which are separate companies hired to perform the assessment. Because each provider has its own methodology and areas of emphasis, one company can be rated best in class by one provider yet worst in class by another. Another inconsistency is the source data a company provides for assessment, which is often just an estimate. In addition, larger, more established corporations tend to score better than smaller companies, which often lack the resources to produce lengthy sustainability reports and get penalized for lack of data.

For example, Tesla's ESG score widely varies with the company that scores them. While some ESG scorers rank them among the best companies for sustainability in the United States, others put them below oil companies and even tobacco icon Marlboro.

The term *ESG* was officially coined in 2005 by UN staff members with the intention of improving transparency and standardizing reporting requirements for companies' sustainability funds, and ultimately reorienting capital toward more sustainable companies. Since then, 186 sustainable exchange-traded funds (ETFs) have been created to give environmentally and socially conscious investors new options for trading. But the lack of standardization has led to accusations of greenwashing, market manipulation, and the promotion of political agendas to the point that certain states have even banned government pension funds from investing in these ETFs. In October 2022, nineteen Republican-led states launched an investigation into the six large US banks that form the Net-Zero Banking Alliance overseen by the UN, which has restricted fossil fuel–related companies from getting loans. Although the investigation hadn't concluded at the time of this writing, the politicization around ESG couldn't be more apparent.

The issue with ESG is that it's too broad, is not transparent, and plays directly into our culture wars. Making progress on taking carbon out of the air is too important, and the environmental aspect of it deserves its own focus. As it is, it acts more like a social justice loyalty oath. For instance, Anheuser-Busch's ESG score plummeted after the company dropped transgender influencer Dylan Mulvaney's brand association after massive controversy. Hate or love Anheuser-Busch's decision, their lower ESG score had nothing to do with the company's focus on the environment. Environmental initiatives succeed when they are focused—and fail when they are broad and feed into the divisiveness tearing our society apart.

Vague sustainability reports, greenwashing, and skewed ESG investing practices are only part of the complicated landscape of how the economy and environmentalism are currently at play. Ironically, oil and gas companies that traditional activists have viewed as the villains in the climate change story are in fact having some of the most positive, trackable impacts on the environment.

I believe a narrowly focused assessment of the progress companies are making to reduce their carbon footprint (relative to their industry) makes sense. For instance, oil and gas companies can do a great deal to reduce their carbon footprint—and the oil and gas companies could be compared to each other for their actions. Natural gas extracted in countries like Russia emits 40 percent more pollutants than gas extracted in the United States. So, highlighting and encouraging the United States' carbon-efficient practices—and potentially onshoring more production—could make a tangible impact. Changes to efficiency, like upgrading technology to lower emissions on natural gas, make a large (and trackable) impact. In general, I prefer the carrot over the stick and hate greenwashing, but I also like metrics and transparency. In short, the concept of ESG has merit, but as it is currently structured, it is not helpful, and I don't believe it is doing much to reduce our carbon footprint. It should be scored independently and not looped in with other issues.

The truth is that, despite the difficulty of navigating the landscape of businesses that claim to be green, having green options at all is still a privileged place to be. As consumers, we can choose to buy a more sustainable product that might cost a little more, with the hope that technological advancement will soon make that product just as cost-effective. And, while there's no guarantee that a product is as ecofriendly as a company claims it to be, ESG and other programs like it are moving in the right direction to give consumers better choices. Whether as customers or employees, it's up to our generation to continue pushing companies forward with the planet in mind—while also demanding transparency surrounding their actual environmental outcomes.

Translating Consumer Demand into Education for the Next Generation of Entrepreneurs

With sustainability becoming increasingly valued by investors, customers, and employees alike, building scalable climate tech businesses will ensure that the US continues to lead in today's industry markets. As our generation enters the workforce, we are at a unique use-it-or-lose-it moment, and universities need to step up their game to keep up with the demand for specialized skills. In a 2020 Cambridge Assessment International Education survey of eleven thousand US students aged thirteen to nineteen, 97 percent indicated a desire to learn about climate change formally in school. Yet, in a different study that year, 40 percent of college students reported having "low or no" coursework in sustainable development.

Since then, colleges and universities throughout the country have begun to seize the opportunity to bolster applicable curricula in environmental entrepreneurship, engineering, and science to meet the growing demand for expertise in these fields. Through a combination of philanthropy and institutional investments, many universities around the country have created schools and expanded coursework to focus on climate change.

For example, in 2021, Columbia University launched the nation's first official climate school to offer programs and research on sustainability, the environment, social justice, and geosciences. That same year, Stanford University announced that a $1.1 billion donation from a single venture capitalist would fund the launch of a new school aimed at researching high-impact solutions to climate change, their first new school in seventy years. Over the next three decades, Yale will funnel $15 million in funding to early-stage projects aimed at fighting climate change. According to the dean of the Yale School of the Environment, the fund will provide support for faculty and students at the School of the Environment who are working on solutions related to all aspects of climate change, ranging from carbon capture to forest restoration to urban development.

Even though the current funding of environmental programs at American universities is unprecedented, meaningful results are not a given. Just as bold promises that companies make about sustainability do not guarantee outcomes, neither does the money funneled toward climate change education ensure that it will be put to the best use. In fact, many university events I speak at, though well-intended, do not go beyond generalized conversation. Very often, I've witnessed little, if any, follow-up after these events.

In 2021, I was asked to speak at the official launch of Duke University's Climate Commitment, a $36 million project that aims to unite its education, research, operations, and public service missions to address four focus areas: energy transformation, climate and community resilience, environmental and climate justice, and data-driven climate solutions. The project is inarguably a noble-sounding one; however, only with time will we be able to gauge its true impact beyond talk.

There are many ways colleges and universities can optimize outcomes, even without huge endowments and showy kumbaya events. Sustainability departments should focus not only on high-level technologies, but also on tactical applications, such as the plausibility of energy sources beyond solar and wind or the ins and outs of green manufacturing. Professors should be teaching practical expertise, fostering open-minded thinking, encouraging entrepreneurship, and supporting research without an agenda. In the next decade, universities need to become the hub of innovation, where faculty, students, researchers, and business leaders collaborate toward finding solutions that can evolve beyond coursework and a degree.

The first land grant colleges in this country were established in 1862 and are a perfect example of this type of collaborative effort with practical, solution-oriented results. These federally funded institutions aimed to teach practical agriculture, science, military science, and engineering in addition to a traditional liberal arts education as a response to the nation's burgeoning industrialization. Today, Texas A&M University is a land-grant college that is home to several programs focused on advancing profitable, environmentally sound farming systems through nationwide research and education grants.

Important projects at Texas A&M include everything from a $4.4 million research project to develop tastier melons, to using the agricultural by-products of corn stubble, grasses, and mesquite to simultaneously produce sustainable bioplastics and biofuel at a low cost, to developing cattle-grazing techniques and crop management to increase soil carbon sequestration.

One of the most fascinating, albeit icky-sounding, projects to come out of Texas A&M is the use of black soldier fly maggots to help take care of food waste, which makes up over 10 percent of global GHG emissions. In a process that takes just two weeks, researchers have figured out how to harness the life-cycle power of these wiggly, wormlike insects to break down food waste. Besides being fast and effective at digesting waste, this process also produces a high-quality fertilizer and a third benefit once the maggots have dried up: high-protein, high-fat feedstock for local chicken farms. Talk about a triple threat to climate change!

Other, more traditional educational institutions, such as my alma mater, the University of Washington (UW), are finding effective ways to interweave important environmental issues into every major, offering coursework in environmental entrepreneurship, environmental engineering, environmental sciences, wildlife management, and climate geology. Being able to apply their specific area of interest in climate change to their major is a game-changer for students at UW and a source of pride for the school's president, Ana Mari Cauce, who is just as passionate as I am about protecting the environment.

UW has been a forerunner in preparing students for the future workforce while also becoming excellent stewards of the planet. To graduate, for example, students are required to take at least four natural world classes, regardless of their majors. In my first *Electric Election Roadtrip* stop, Ana and I visited Friday Harbor Laboratories, a UW marine biology field station where students and scientists are devoted to researching important environmental topics such as ocean acidification and its effect on local marine life. Another initiative of UW is their multidisciplinary climate impacts group, which focuses not just on atmospheric research but also on how their findings are communicated so that the public understands the complexities of climate change.

In recent years, more and more colleges throughout the country, including UW, are offering case competitions specifically having to do with climate change solutions. These *Shark Tank*–like events are opportunities for students to compete before a panel of local business leaders for university grants that will help them continue to fund one-of-a-kind projects. From the installation of composting bins everywhere on campus to making EVs available for students to rent, UW's leadership is ensuring that students are not just learning in the classroom, but also experiencing environmental stewardship in their daily lives.

A Word about Nuclear

As the world searches for clean energy solutions, one field of study that has regained significant momentum over the last several years is nuclear engineering. North Carolina State University in Raleigh is home to the nation's first non-governmental nuclear reactor and is a leading nuclear research center geared toward making nuclear energy safer, cheaper, and more reliable. Since 2020, enrollment in nuclear engineering programs at NC State has nearly quadrupled, with many graduates ending up working for the United States Department of Energy or private nuclear energy companies.

As our domestic nuclear energy industry continues to lead the way for the rest of the world, it will play an important role in job creation and economic growth. At the time of this writing, the US nuclear industry already supports nearly half a million American jobs and contributes about $60 billion to the US Gross

Domestic Product each year. The US nuclear industry also has the highest-paying jobs in the entire electric power generation sector, with salaries that are 30 percent higher than the average local profession and up to 25 percent more per hour than the next best-paying electricity-related job.

IT MAY SURPRISE YOU TO KNOW . . .

The US is the world's largest producer of nuclear power, accounting for more than 30 percent of the worldwide nuclear generation of electricity. Nuclear energy already makes up 20 percent of the US electricity supply, which is over half of our zero-carbon electricity. This means that if we were to double our nuclear output to replace other electricity sources, then all of our electricity would be clean.

Although nuclear energy still remains a heated topic for many, it is important to note how far *fission* technology has come since its inception. The first generation of nuclear reactors were built inside of large government facilities as a single product that had to last for fifty to sixty years. Nowadays, much of nuclear energy has shifted to the private sector, where smaller reactor designs and exchangeable parts have significantly driven down costs and production time.

Investors have caught on to the scalability of these modernized versions and are heavily engaged in trying to bring commercial *fusion* reactors to market. Billionaires such as Bill Gates and

Jeff Bezos have recently contributed nearly $2 billion to these projects and are poised to fund more as innovation continues. Although the federal government has spent about $600 million annually on fusion research and development, many startups are eschewing government grants to raise smaller, venture capital–backed funding rounds and move through the approval process much more quickly than their predecessors.

Oklo is a fast-growing California-based startup that has relied solely on private investments for its first reactor, dubbed the Aurora. Housed in a minimalistic A-frame building hundreds of times smaller than traditional, old-school reactors, Aurora will, once built, run on used fuel recovered from an experimental reactor in Idaho that has been shut down since 1994. Oklo anticipates that the reactor will not need to refuel for another twenty years. This type of efficiency is a far cry from fusion reactors of yesteryear and an exciting indication of where the future of nuclear energy is heading.

Nuclear is only one of many clean energy solutions on the cutting edge of climate tech that depend on young entrepreneurs to step up and fill in the innovation gap. As a long-term solution, it holds the potential to fuel the future safely, affordably, and cleanly.

While environmental activists often balk at the ability of capitalism or the marketplace to solve climate change, it's the best economic system we have to solve these challenges. Without profit-driven incentives to develop innovative technologies that can stand on their own merit, we'll never uncover solutions that decrease costs and increase efficiency for the average person. And if the average person can't adopt the solution, the solution is pointless.

As we've covered, the countries with the freest economies in the world have the best track record of reducing emissions over the past few decades. At the end of the day, the government cannot regulate itself out of climate change.

Now let's take a look at some lower-hanging fruit that can provide immediate short- and long-term answers . . . literally right in our backyards.

The 411 from Chapter 5

- A competitive spirit to solving climate change spurs the most innovative ideas by providing people with "skin in the game" and decreasing costs for technologies to stand on their own.

- America has been leading the way in emissions reductions, but we need far more entrepreneurs to help uncover solutions.

- The bipartisan Energy Act of 2020 and incentives from the Inflation Reduction Act were a start in advancing carbon capture, emissions reduction in coal and gas, and nuclear development, and supporting climate tech businesses.

- Greenwashing and ESG scoring are challenges to the green economy, but also provide a great opportunity for improvement.

- Universities and young people have a massive role to play in closing the "innovation gap" that exists on climate change.

6

Conserving Our Environment, Our Heritage, and Our Future

We all remember *The Lorax*, the movie about a world in which the Truffula trees disappear at the hands of the Once-ler, a greedy business magnate. The story, based loosely on a Dr. Seuss classic, delivers the right message to its young viewers about caring for the environment, but the wrong one about how to go about it: leave the forest alone to grow again on its own. It's an oversimplified solution to a complex reality, an approach that is at the center of every preservationist's thinking and, admittedly, one I also shared as a small child.

As a kid who had watched *The Lorax* and who grew up surrounded by the Northwoods of Wisconsin, it was hard not to look at lumber workers and hunters as the enemies of nature as I watched them take down the trees and animals I loved so dearly.

And I don't use the word *love* lightly. As a self-proclaimed vigilante of all creatures great and small, I couldn't bear to see even the smallest of insects suffer. I did my best to rescue them from harm—even to the point of removing a fly or a beetle from the sticky filaments of a spider's web. Little did I realize that some of these heroic moves might actually be disrupting a natural sequence of events (including a hard-earned meal for poor Charlotte). But who could blame me at age five?

Now, as a conservationist, I recognize that protecting the environment is not as black-and-white as preservationists—or kindergartners—may think. Since the time of cavemen, humans have, to some degree or other, had a necessary impact on nature, whether it's burning wood to stay warm or hunting and fishing to stay fed. Over the past several years, I have visited some of the most pristine landscapes across the country, as well as some of the most devastated ones, and I am well aware of the negative impact we humans have had on the planet. Yet instead of denying our continued reliance on the Earth's resources, I've found that solutions to help mitigate that impact on the planet also benefit the people who inhabit it.

The question of how best to care for the environment is not a new one. In fact, it has been a source of great debate in our country since our early pioneer days. Although John Muir and Gifford Pinchot were both nineteenth-century nature lovers, immortalized for their deep devotion to protecting the great outdoors, they didn't always agree on how to go about it. The story goes that one day the two men were walking together along the rim of the Grand Canyon and, coming across a tarantula, Muir stopped

Pinchot from killing it, saying, "It has just as much right to be here as we do." Pinchot wasn't sold on that idea.

Muir, founder of the Sierra Club and co-creator of important protected wilderness areas, including Yosemite National Park, was a dyed-in-the-wool preservationist. He believed that nature should be left alone, untouched by foresters and other so-called meddlers. Pinchot, on the other hand, as the first head of the US Forest Service, was outspoken about conservation and the important uses of public lands, including grazing, agriculture, and lumbering.

Both men loved nature. However, in today's world of climate change and disastrous events such as ever-increasing wildfires, the stakes are too high to leave to philosophical discussion. The urgency with which the world needs to act requires looking beyond partisan agendas and alarmist calls to action and finding solutions that work now.

As I've already said, climate change is not a political issue—it's a human one, and it's no secret that we humans are now paying the price for the substantial harm we've done to our planet. However, it's a much lesser-known fact that one of the major contributors to climate change is poor land management and the same hands-off approach that environmental propaganda such as *The Lorax* relentlessly promotes. It is true that a loss of the world's forests accounts for 8–10 percent of global CO_2 emissions increases, and that reversing this trend is an important component in reducing emissions. After all, forests around the world absorb roughly 7.6 *billion* metric tons of carbon annually, a carbon sink equivalent to 150 percent of the United States' annual emissions. Yet deforestation is only part of the problem.

When I was a college student, I used to dread the first hour of the drive from UW to Crystal Mountain ski resort, which took me through a never-ending landscape of clearcut forest. I assumed that this depressing scenery was the sad aftermath of overlogging or diseased trees. Only later did I learn that it was actually the result of an important forest-thinning practice used to prevent wildfires. I was even more relieved to learn that the wood from the felled trees was being used for heating and building materials to benefit the surrounding local communities.

It was one of my first introductions to forest management and an eye-opener for me, especially because for months I'd been hearing news stories about the devastating forest fires that were happening in the west because of climate change. I remember looking out of my fraternity house window one morning onto a hazy street of soot-covered cars and a bright orange sky in the background and thinking, *Holy Crap! Climate change really is burning up our planet!* I'd been so used to Washington's Governor Inslee blaming these fires on global warming that I'd hardly questioned the truth behind those claims.

I started researching the data and speaking with rangers and lumber workers to find out the truth about the health of America's forests. Seeing the statistics was a big wakeup call. I realized how much forest health impacts climate change—a lot more so than the other way around, as Inslee and all the news syndicates were purporting. In 2020 alone, California forest fires emitted more greenhouse gases than any reductions that state had made in twenty years. That's twenty years of effort completely gone to waste! The World Economic Forum also reported that in 2021, nearly 2 billion tons of carbon were emitted from forest fires, equivalent to Germany's total annual CO_2 emissions. The 7.6 billion metric tons of carbon forests soak up are overshadowed by these fire emissions. We're screwing up pretty badly if our most important natural carbon sink—forests—are a net-negative on our environment.

More jaw-dropping than these numbers is the fact that they aren't even being calculated into overall global emissions. Apparently, because the experts consider forest fires *natural* and *unpreventable*, they somehow don't count. *Natural* and *unpreventable*? With 85 percent of US wildland fires caused by the results of human activities such as unattended campfires, discarded cigarettes, equipment use and malfunction, and arson, the idea is nothing more than a blatant untruth. And they definitely count when figuring out how much progress we are making toward our 2050 world goal of zero emissions.

Instead of being upfront about the true cause-and-effect factors in climate change, Governor Inslee and other figureheads like him continue to pay lip service to the environment, billing themselves as climate change warriors in order to push a very specific socio-economic agenda. After all, touting EVs and clean energy as the cure-alls of climate change is much more appealing to urban constituents than talking about tree-thinning practices and wetlands restoration. Likewise, blaming climate change as the cause of forest fires is an excellent strategy to incite more fear, motivate environmentally conscious consumers, and distract voters from the real solutions that are right under our noses.

Yet the historic proportion of the Canadian forest fires of the early summer of 2023 was impossible to ignore as wind carried their smoke into cities as far south as Atlanta, Georgia. More than 150,000 Canadians were displaced from their homes, over 8.8 million hectares of forest were destroyed, and the reparation costs were estimated in the tens of billions of dollars. Three hundred is considered a hazardous level on the Air Quality Index (AQI). Cities such as New York City reached 413, causing school closures and public health warnings. The summer before the Quebec fires, Europe also experienced an unprecedented number of fires and contributed largely to the 1,455-megaton increase in worldwide carbon emissions in 2022. Yikes!

Although strategies such as increasing renewables and lowering our dependence on fossil fuels are vital to our long-term fight against climate change, they require enormous amounts of investment, diplomacy, and time. Natural climate solutions, on the other hand, can provide some of the most immediate, cost-effective, short-term solutions to reduce greenhouse gas emissions. These include land restoration, forest management, sustainable agricultural practices, and food waste mitigation, and all help nature continue to do what it's been doing for millions of years: capturing and sequestering carbon. They also offer huge economic opportunities, leverage private-sector investment, engage rural communities, and gain significant bipartisan support. Last, but not least, nature-based solutions are the only way to conserve some of America's most beautiful landscapes and ecosystems, which are in danger of being lost forever.

If you're like me, for you the great outdoors is a place to find healing and peace of mind, a place where you can relax and recuperate from all the stresses of life. After college, I moved to the waterfront neighborhood of Seattle called Ballard, where whenever I felt stressed or wound up, a quick run to the ocean would change my entire perspective, and I'd feel grounded once again. Hearing the cries of seagulls, the barking of seals, the gentle roar of waves brushing against the shore, seeing the snowy mountain peaks in the distance—all of this reminded me that life's daily problems are not all that big. For me, nature has always had this magical power to soothe. Protecting this magic is the *why* of what I do. I'm sure you have a similar *why*.

Sadly, for many people, nature has turned into a political battleground where the sounds of birds and rustling leaves can hardly be heard over the noisy debates about climate change. Like divorced parents who are too busy arguing over custody to see what's best for the child, politicians on both sides seem to have lost their personal connection to nature (if they even had one to begin with) because they are so preoccupied with gaining support for their own agendas. If we aren't careful, our generation will do the same and will, quite literally, no longer see the forest for the trees.

Natural climate solutions are a strategy that both sides of the political spectrum can come together on. Bipartisan bills such as the Growing Climate Solutions Act, which promotes sustainable farming practices—with a ninety-two to eight win, including forty-five Republican votes—prove it.

By and large, the GND does not focus on any of the powerful nature-based solutions I'm about to discuss that could unlock massive GHG reductions and at the same time restore our precious landscape. Current data shows that conservation practices have the potential to account for 37 percent of all emissions reductions required by the Intergovernmental Panel on Climate Change's goal of halving global emissions by 2030. Take Alaska, where the average fire management cost of a fire season is approximately $133 million. If, however, that state spent 1 percent more on correct forest management per year, it could be able to reduce its forest fire emissions by a whopping 40 percent.

Taking the Politics out of Our Forests

American conservatives have always been natural conservationists—it's in our name after all—and our generation intuitively knows that nature-based solutions are a powerful way to reduce emissions. In a recent ACC poll of 1,200 adults aged eighteen to thirty, 79 percent agreed that planting trees and restoring ecosystems was an effective way of fighting climate change, and was in fact the most popular solution of all in the survey. Fifty-eight percent found carbon capture and sequestration equally important.

Carbon sequestration is an important new endeavor trending in the climate tech world. Startups as well as multibillion-dollar corporations are racing to develop systems that will essentially suck carbon out of the atmosphere and store it below ground indefinitely. While this technology is useful in the climate action toolbelt, many in the environmentally conscious community have overlooked the unique carbon-sequestration capacity of a tool the Earth already possesses in abundance: trees.

In essence, a tree is one giant storehouse of carbon that tenaciously holds on to the element with every one of its parts, including its trunk, branches, leaves, and roots, even after it's been harvested. In one year, a mature, live tree can absorb more than forty-eight pounds of carbon dioxide, which is permanently stored in its fibers. In 2018 alone, global tree cover removed 37.1 metric tons of CO_2 from the atmosphere, and the US's 766 million acres of forestland continues to remove nearly 16 percent of American emissions each year. Now, that is the most powerful carbon sequestration technology I've ever seen! Yet, despite being an excellent frontline defense weapon against global warming, trees receive little or no recognition in the conversation about climate change solutions.

Sure, activists believe in the importance of trees, but their tactics accomplish nothing to protect the long-term health of our forests and woodlands. We've all seen and heard about protesters having standoffs with chainsaw arms or even tying themselves to tree trunks to stop loggers from chopping trees. It's the kind of story that incites fear and rage, which in turn, well, incites more fear and rage. Such theatrical displays start and end at the exact same place, with no real progress to speak of in between. Still, that's the stuff that makes headlines.

Not many Americans know about the powerful policymaking going on simultaneously behind the scenes. The bipartisan Save Our Sequoias Act, introduced to Congress by Republican congressmen Kevin McCarthy and Bruce Westerman, aims at partnering with state, local, tribal, and private entities to improve forest-fire resilience and enhance reforestation and rehabilitation projects. I've spent a lot of time in Washington and have seen the challenge politicians face when a policy they believe in falls to the wayside simply because not enough constituents know about it. Truth be told, a bill like the Save Our Sequoias Act has very little clickbait value and can easily get pushed to the back of the line without enough public support.

In addition to sound behind-the-scenes policymaking, some states are doing a much better job than others in collaborating on effective climate action. Florida has a long history of bipartisan environmental policy that works. And several of the state's public-private partnerships can serve as a national model to work across party lines to get things done. Nearly two-thirds of southeastern Florida's forests are privately owned, a percentage much larger than that of western states, which happen to have some of the poorest land management records, simply due to a lack of resources, infrastructure, and incentives.

During the Electric Election Roadtrip, I spoke with Julie Wraithmell, vice president and executive director, Florida, of the National Audubon Society, and Michael Dooner, president and founder of Southern Forestry Consultants, nearly two years after Hurricane Michael tore up much of the coastline of Florida's central panhandle. Entire forests still lay flat from the eyewall of the storm, representing not just damaged ecosystems, but the devastated property of many private landowners who had always viewed the timber that grew on their forestland as a retirement investment. In a matter of hours, that investment had been completely leveled. The increased frequency and intensity of such extreme weather events has left thousands of coastal Florida residents wondering whether to reinvest in their land and restore these damaged forests.

State and local nonprofits such as the Audubon Society and Southern Forestry Consultants understand the importance of public-private partnership in regions susceptible to storms and are committed to supporting conservation efforts that cross partisan lines. Still waiting for federal recovery funding, Floridian private landowners have leaned on the support of these and other public entities to start to restore their properties, even with a lack of government money flowing in.

Forty- or sixty-acre properties may seem like small change in the climate action world, yet in the case of private property, the sum is often greater than the whole. Private landowners have a serious stake in seeing their properties restored, even though these are not people who necessarily call themselves environmentalists. Nature-based solutions present an all-around win for individuals, communities, businesses, and the planet. It's climate action at its greenest, both economically and environmentally, and therefore, no added incentivization is needed to get the job done.

As Julie Wraithmell says, "Floridians live closer to the land than a lot of the country does, and by that I don't just mean that we're rural, because there's a big part of Florida that's urban. But what I mean is the environment's our economy. If that goes south, everyone suffers, whether you're an outdoor person or an AOC devotee. And so, we have a long history in Florida of the environment not being a partisan issue." Indeed, the work being done by Florida governor Ron DeSantis, the state house, and the state senate across party lines would have immeasurably positive results if it were scaled nationally.

The Tipping Point of Trees

For generations, the image of Smokey Bear has greeted Americans in public parks across the country to warn against the dangers of human-caused wildfires. This adorable icon of the US Forest Service is part of the longest-running public service announcement campaign in our history. The campaign has even managed to keep up with the times with a mobile app with features that include a step-by-step guide to campfire safety and a map of current wildfires across America.

While I love Smokey Bear—and who doesn't?—far more hands-on work needs to be done to save our forests and fight climate change. One recent study shows that wildfires in boreal North America could, by mid-century, cumulatively contribute to nearly twelve gigatons of carbon dioxide emissions. That's about 3 percent of the remaining global carbon dioxide emissions associated with keeping temperatures within the Paris Agreement's 1.5 degrees Celsius limit.

When it comes to our nation's forests, being hands-off backfires—literally. Unpruned forests can easily move past a delicate tipping point and transform from carbon sinks to carbon emitters. Stand density, the number of trees within a specific area, plays a large role in fire growth. America's forests are far thicker than they were just one hundred years ago. Back in 1911 in one California forest, for example, foresters recorded an average of just nineteen trees larger than six inches in diameter per acre. In 2013 in that same forest, 260 such trees stood per acre, acting as a virtual tinderbox, just ready to be lit. Ground fires in such conditions can generate enough heat and intensity to move quickly from the forest floor up into the tops of the trees to become what's known as crown fires, which at that point are nearly impossible to control.

Research shows that proactively managing forests can be one of the best fire-preventative tools around. Selective harvesting, thinning treatments, brush removal, and pruning are all practices used by foresters to thin out forests so that less fuel is available once a fire is ignited. Controlled burning in what's known as *prescribed fires* can be one of the most effective methods to reduce the risk of catastrophic wildfire because it creates gaps among trees as natural borders to contain the flames. All of these active fire-management tools can make forest fires more manageable and reduce emissions when wildfires do inevitably occur.

As I've said often, people care most about the environment in which they live. And the level of government closest to these people is the most effective at implementing policies that promote conservation. The federal government has shown us that it is unable to keep up with caring for the vast amount of the land it owns, which is about one-third of America's total geography.

Growing up, I visited several national parks with my family each summer and watched from the backseat of the car as my dad happily handed a twenty-dollar bill to the entrance booth clerk. He always told us he didn't mind paying the fee because the money would go back into the upkeep of the place. Little did he or any of us know at the time that this was not the case. In fact, until 2019, when President Trump signed an important bipartisan bill called the Great American Outdoors Act, which ACC helped pass, national parks were over $13.6 billion in debt. Even though the act has helped to redistribute entrance and camping fees back to the national parks, the lack of resources for forest and land management poses a serious problem to our nation's federally owned land.

When possible, we need to put land stewardship in the hands of local governments and private landowners. Green Diamond Resource Company is a family-run timber company at the forefront of forest stewardship practices. With timberlands in California, Oregon, Washington, Montana, and the Southeast, the company understands the unique ecology and economy of each of these regions and works with the latest technologies and conservation techniques to ensure that only 2 percent of their trees are harvested per year.

With sustainability baked into the business, almost all of their efforts are spent cultivating a rich forestland that will thrive for generations to come. Working with state and federal agencies, the company has even signed several voluntary habitat conservation agreements in return for lumbering rights for the next fifty years.

Grasslands and Wetlands: The Forgotten Children among America's Natural Carbon Sinks

Once upon a time, 50 percent of the eastern US was covered in beautiful, bountiful grasslands, where prairie dogs and bison roamed freely amidst the billowing needlegrass and sunflowers.

If this sounds unbelievably picturesque, it was. Sadly, over the past century, 99 percent of US grasslands—and 20 percent of the entire world's total grasslands—have disappeared. Most people I speak to don't even know what grasslands are because they have become so scarce, along with the species that inhabit them, yet besides being one of nature's geographic gems, they are also one of the most efficient ecosystems for capturing and storing carbon.

What exactly is a grassland? Any area dominated by non-woody vegetation such as prairie plants and tall grasses. (Think *Little House on the Prairie* or where you would picture Punxsutawney Phil laying out his welcome mat.) They cover approximately 25 percent of the Earth's land surface and contain roughly 12 percent of the terrestrial carbon stocks on the planet. The fibrous root grasses that are native to grasslands are what make their soil such an excellent sponge for what's known as soil organic carbon (SOC).

Grasslands have agricultural benefits as well because they promote biodiversity, support pollinators such as bees and butterflies, and host predators that can help suppress potential pests. Unfortunately, the conversion of grasslands into farmland significantly reduces the below-ground biomass in roots and minimizes SOC retention. In addition, some conservationists' overzealousness for planting trees in places where forests historically have never existed has destroyed much of the rich biodiversity of these regions.

Afforestation is the practice of planting forests in areas where trees previously didn't grow. In April 2022, I toured the Couchville Cedar Glade State Natural Area, one of the only grasslands left in the state of Tennessee. While there, I learned that such practices are one of the main reasons grasslands in this country have all but disappeared, taking with them immeasurable amounts of native wildlife. I saw firsthand that the feel-good mindset of "Let's just plant more trees" makes no sense when entire populations of species, such as the monarch butterfly and the bobwhite quail, are collapsing as a result.

Our tree-centric thinking might please the Lorax and real-life preservationists; however, it has really only served to trade in one problem for another. This scenario reminds of yet another Dr. Seuss book, *The Cat in the Hat Comes Back*, in which the mischievous title character promises to remove a pink stain around the bathtub, only to spread it throughout the entire house, creating an uncontainable mess that is far worse than the original problem. Similarly, the lesson to be learned with poorly planned afforestation is that when we carelessly continue to pursue an idea that initially sounded good but turns out to be unsound, the chaos that ensues can have serious environmental repercussions.

YOU MAY BE SURPRISED TO KNOW . . .

Besides bad afforestation and overdevelopment of buildings and highways, one of the biggest culprits of the destruction of grasslands has been the overproduction of corn. What might surprise you is that this corn is not the kind for popping or roasting, but instead is bioengineered to be used for fuel. In 2007, Congress enacted a federal mandate to require vehicle fuel makers to blend ethanol into gasoline in an effort to reduce GHG emissions and increase energy security by reducing US dependence on Middle Eastern oil. As a result, as much as 40 percent of the corn now grown in this country is used to make ethanol.

At the time, the plan to mass-produce ethanol may have seemed like a good move for the planet; however, we soon learned that, as with every climate change solution, a series of subsequent trade-offs was inevitable. As farmers plowed up grasslands to grow corn, they released the carbon that had been stored in the soil for decades as well as high concentrations of nitrates that wound up polluting waterways in places such as Des Moines, Iowa.

Today, many lobbyists have been recklessly pushing to triple the amount of ethanol American fuel makers put into gasoline, arguing that using more of this so-called renewable fuel will benefit the environment. In reality, more ethanol means more carbon emissions, more algae blooms, and higher water bills for Midwestern residents.

Replanting grasslands around highways and powerline areas where soil is fallow and unused is a fast, relatively easy solution to restoring this vital habitat. Roadside enhancement projects in former prairie states such as Iowa, which has about 600,000 acres of unused roadside habitat, make complete sense. Grassland plants require little maintenance, beautify the landscape, and help support small wildlife, all while reducing GHG emissions. Such projects are already underway and have achieved relatively easy success in England, parts of Africa, and India.

Like our grasslands, US wetlands, home to waterfowl and other prized game, are some of the most valuable carbon-sequestering areas in the country. According to Restore America's Estuaries, coastal landscapes have the capacity to hold up to 233 million tons of CO_2 per year, and have carbon sequestration rates that are ten times higher than those of forests on an acre-for-acre basis. To put this into perspective, losing 2.5 acres of coastal wetlands releases the same amount of carbon as losing 25 acres of native forest.

Unfortunately, America's coastlines have seen vast amounts of damage due to the frequency and severity of recent storms. The historically rich city Bayou La Batre, Alabama's seafood capital, is another southeastern coastal community I visited, where I learned about how far public-private partnership can go in restoring these areas. The Lightning Point Restoration Project, a collaboration between Mobile County, the Alabama Department of Conservation and Natural Resources, and other conservancy programs, has been thoughtfully restoring and enhancing important waterfront habitats for the past several years.

A far cry from simply planting a bunch of trees or dumping sand dunes onto the shore, these restoration projects are specifically designed and engineered to mimic nature's intelligence. For example, to protect the shoreline, 1.5 miles of overlapping breakwaters were constructed on both sides of the navigation channel to provide an important buffer from waves and boat wakes.

For any of these conservation efforts to work, the Lightning Point Restoration Project needed buy-in from the governor, mayor, and local community members who were instrumental in the decision-making every step of the way. It's important to note that many of the project's greatest supporters are not your typical environmentalists. They are simply natural stakeholders, many of whose livelihoods rely on the health of their waterways and shorelines to commercially fish and build ships.

Not far from Bayou La Batre is Everglades National Park, a place I frequented as a kid whenever we visited my maternal grandmother in southern Florida. I remember the fascination and awe I felt touring these wetlands by boat, as I spotted alligators, crocodiles, and exotic birds lingering on the banks. A few years ago, I toured this rich ecosystem once again, and this time, rediscovered it in a brand-new way. I learned from my guide from The Everglades Foundation about the region's unique role as a buffer to damage caused by rising sea levels, as well as its natural water-filtering properties, and, most importantly, its ability to lower GHG emissions.

Throughout the Everglades' 1.5 million acres of connected wetlands, a blanket of organic peat lies on top of bedrock. This peat acts as a giant carbon sink—that is, as long as it remains wet. Because development in recent years has cut off hydration from Lake Okeechobee, their original water source to the north, much of these wetlands now tends to dry up during the year, causing a release of this carbon, and even destructive wildfires. As part of a comprehensive restoration plan to enhance carbon sequestration in the area, the construction of a nearby reservoir is already in the works.

While this and other similar conservation projects might not seem as glamorous as planting a million trees or installing a million solar panels, they promise to yield results, the sum of which far surpasses their parts. Not only are they guaranteed to lower emissions, but such efforts are the only way to restore the natural beauty of our country before it is lost forever.

Nature-Based Solutions Beyond Our Coastlines

Just as we have a responsibility to be good stewards of our own land, the United States has a duty to lead the world in leveraging natural solutions. It's not enough to be able to boast of lowered GHG emissions and conservation efforts on our own soil if we are continuing to do damage on other continents.

Ecosystems around the world act as significant carbon sinks with global significance for the fight against climate change. Take Brazil's Amazon region, often dubbed *the lungs of the Earth.* This sprawling rainforest is responsible for reabsorbing nearly a quarter of all carbon dioxide in the Earth's atmosphere. The Amazon also has a critical cooling effect on the planet as its trees channel heat away from the Earth's surface and move it high into the atmosphere.

Yet for years, commercial farmers from all over the world, including the US and China, have been buying up acreage and cutting down trees to use the land for cattle ranching and high-demand crops such as soybeans. To date, about two thousand square miles of the Amazon forest are being destroyed per year by foreign agriculture. To make matters worse, the cleared-out forest is poorquality farmland that isn't even usable after a few harvest seasons.

What was once the Earth's precious, natural protector from GHG is being treated as a disposable resource and quickly becoming a veritable wasteland. Yet who in Washington is talking about this?

The sad truth is too many environmental activists are not even aware of what's happening beyond their backyards, much less beyond US borders. Many live in urban or suburban settings that are far removed from the nature that is being exploited and poorly managed. They are more preoccupied with increasing the number of EV chargers in their neighborhood than they are with restoring the world's forests and oceans, proving that *pro-climate* does not necessarily mean *pro-environment*. It's a disconnect that is costing the planet its natural abundance and beauty.

Yet there is a group of people who actually do feel close to the land and have a vested interest in its future. Each year, forty-seven million American hunters and anglers head into the wilderness, investing hundreds of millions of dollars of daily expenditures to support jobs in every corner of the country, from gear shops, to gas stations, to coffee shops, and strengthen the American economy to the tune of $200 billion per year. More importantly, a large portion of the revenue from their license fees and federal excise taxes on firearms, ammunition, rods, tackle, and even motorboat fuel goes back into funding for public land management.

Understandably, among preservationists, hunting and fishing are activities that have always received a bad rap. It probably comes as no surprise that during my early bug-saving days, I found the deer hunting that was so popular around my hometown both cruel and unnecessary. Of course, as an adult, now that I understand the concept of restoring balance to nature, I see that with fewer predators roaming the Earth, overpopulation of wildlife such as deer can pose a serious threat to many valuable ecosystems. Historically, hunting and fishing have helped to regulate these populations.

President Theodore Roosevelt, an avid hunter and fisherman himself, made conserving America's land resources a national priority over a century ago. Within a little more than a year of taking office, he had established two additional National Parks, Crater Lake in Oregon and Wind Cave in South Dakota. In 1903 he set up Pelican Island, our first federal bird reservation and the beginning of what would become the National Wildlife Refuge System. Altogether, Roosevelt left behind a legacy of 230 million acres of federal land for conservation and doubled the number of National Park sites.

You may remember my discussion in Chapter Four of Ducks Unlimited, an organization that has become a world leader in wetlands and waterfowl conservation. Since its founding in 1937 by a small group of hunters, DU has managed to conserve more than fifteen million acres of habitat across North America in the areas that are most important to ducks and geese. To date, the organization has reforested more than 178,000 acres in the Mississippi Alluvial Valley alone, an area previously cleared for agriculture and other purposes, and worked to restore backwater to these forests to mimic historical flooding. Ninety percent of its current membership is still composed of hunters, and most of its collaborators are farmers, ranchers, and other landowners, all of whom have natural stakes in the health of the land.

America is not the only place where hunting and conservation can go hand in hand. Across the ocean, trophy hunting, the legal shooting of game under a regulated official government license, has become a booming industry in Africa, where hunters are allowed to kill carefully selected species of elephants, lions, and rhinos, and more with restrictions on location and weaponry that can be used. Surprisingly, it is an excellent way to preserve species of animals that are endangered or near endangerment because it guarantees protection of their land where habitats have been rapidly converted to crop farms and livestock ranches. This loss of habitat, as I've already talked about in the Amazon, is one of the greatest threats to precious wildlife as more and more farmers and ranchers are clearing vital woodlands and killing wildlife that pose a threat to their livestock or compete for forage.

In South Africa, most hunting reserves are former cattle ranches that were converted back to wildlife habitats after legal changes gave landowners incentives to maintain high-quality habitat on their land. As an added incentive, landowners also have the opportunity to sell hunting rights on their land. The revenues procured from trophy hunting get put back into the land to ensure continued conservation. In addition, in countries such as Tanzania, expenses related to the government's anti-poaching enforcement are funded by trophy-hunting revenues. (Poaching, by the way, is illegal hunting, which for decades has been a rampant problem throughout Africa.)

Legal trophy-hunting revenue does even more than just help wildlife and the environment. It also goes toward impoverished communities through cost-sharing arrangements with trophy-hunting operations under which community members receive half of the accumulated revenues each year. This extra profit has led to a 15–25 percent increase in household incomes in these areas.

Climate Change and the Big Blue

Between America and Africa, of course, lies the Atlantic Ocean, an enormous expanse of blue carbon, the term for carbon captured by the world's ocean and coastal ecosystems. If a silver bullet did exist for emissions reduction, then blue carbon might be it. Yet, besides the imminent threat of "sea levels rising," how often do you even hear about the ocean in terms of climate change solutions? My guess is hardly ever, even though our coastal and marine ecosystems have the potential to naturally sequester about 25 percent of the world's emissions. In fact, seagrasses and mangroves, along with their associated food webs, can absorb carbon dioxide from the atmosphere at rates up to four times higher than terrestrial forests can. They also support healthy fisheries, improve water quality, and provide coastal protection against floods and storms.

Sadly, well over 50 percent of marine ecosystems have already been lost to the effects of global warming, acidification, and disease. Not only is this loss tragic from a nature lover's perspective, it also, if it continues at the current rate, could release up to one billion metric tons of CO_2 per year, adding to, rather than mitigating, climate change. Similar to our forests, the world's greatest natural carbon sink could become one of our biggest emitters once it reaches its imminent tipping point.

Coral reefs, although they now cover less than 0.1 percent of the world's oceans, support over 25 percent of marine biodiversity and offer sources of medicine, tourism revenues, and natural coastal protection that ultimately amounts to enormous government savings when it comes to storm recovery. It is estimated that one foot of coral reef, in fact, can save approximately $1 billion in storm damage.

The Florida Reef, which stretches approximately 350 miles from the Dry Tortugas to the St. Lucie Inlet, has lost all but 2 percent of its living coral cover in recent decades. The good news is that many efforts to protect and restore blue carbon are underway. In the late fall of 2022, I visited Mote Marine Laboratory in the Florida Keys, where scientists had just reached the exciting milestone of restoring one hundred thousand corals, which are now actually starting to regenerate on their own. With extensive land-based and underwater nurseries, another fifty-six thousand or so coral fragments are on their way to replenishing several species of lost coral. The lab has met with such great success that they believe they will be able to speed up the coral regeneration process to take only a few years instead of the originally estimated twenty-five years.

In an effort to reduce shipping emissions to zero by 2050, we can start with short-term measures such as slow steaming (operating ships at below-maximum speeds), as well as long-term changes, including switching to zero-carbon propulsion systems. Green shipping corridors that support zero-emissions technologies and sustainable alternative fuels for ships are another way to help with the decarbonization of our oceans. Two such corridors already exist between Los Angeles and Shanghai, and between Antwerp and Montreal.

While offshore efforts to restore blue carbon are important, the problem of much of our ocean's pollution starts on land as the runoff from fertilizers, detergents, and even sewage eventually flows into lakes and seas. The result is excessive algal and microbial growth that has led to at least five hundred sites in coastal waters being classified as *dead zones*, areas with so little oxygen that most marine life cannot survive. The largest of these, in the Gulf of Mexico, is approximately 4,280 square miles. It may come as no surprise that agriculture contributes largely to this runoff.

Many farms, especially those belonging to the cattle industry, are actively lowering their emissions each year, which still accounts for 11.2 percent of our country's emissions. As we saw, startup companies such as Brightmark in Wisconsin are already helping to convert cow and swine waste to clean energy. By running this methane-intensive manure through an anaerobic digestor, the company has found a way to create a renewable biogas that's as effective as any natural gas out there, with a compostable fertilizer by-product and zero emissions.

By the way, Brightmark also has an inorganic division that diverts plastic from landfills and instead, using their own proprietary technology, generates three types of by-products: a low-sulfur diesel, a blend-stock used to make new plastics, and wax to make candles or coatings for the inside of cardboard boxes. This no-waste production line is conservation and innovation at their very best.

Zeina El-Azzi, the chief development officer of the company

and a true advocate of public-private partnership, has described Brightmark's dynamic working relationships with dozens of local farms. In her words, "Brightmark isn't just a developer. We're an owner and operator of these assets and we're here for the long haul, so if something goes wrong or breaks down, we're going to be there to fix it . . . When we contract to take the manure from dairies, we are guaranteeing a minimum payment to them every single month or quarter of every single year." Zeina also described how government incentives can help build more projects similar to Brightmark. Programs such as the low-carbon fuel standard in California offer credits to offset costs for building biofuel projects like it.

Still, wind turbines and solar panels are more appealing to talk about than manure, and many clean energy startups need help getting the word out there about what they're doing. You may want to tell your local representative about companies like Brightmark. You may also want to share the videos that Brightmark posts on Instagram that teach about the poop-to-power magic that's happening on farms across the country. Although the hope is that these types of solutions become international in scope, experience tells us that they first need to start with support from our own backyards.

Besides the *udder*ly—I couldn't resist the pun—impressive resources I've just talked about that are working with nature-based climate solutions, improvements in feed and production are also driving down the 4 percent of total US emissions that our country's cattle are responsible for. In the last five years, US farmers and ranchers in general have been joining in on climate action and have installed 132 percent more renewable energy sources, including geothermal, solar panels, windmills, hydro systems, and methane digesters.

While these breakthroughs in integrating new technologies are important, global agricultural emissions still contribute majorly to climate change, generating 19–29 percent of GHG per year.

It would be easy to say that as long as farmers continue to embrace important nature-based solutions to climate change, farms will help lower this number significantly and also become more efficient. However, in truth, farmers will always care first and foremost about the efficiency of their farms because, after all, their livelihoods depend on it. The effects on the environment are also important to them, but they are less immediate and therefore secondary. Consequently, the more accurate way to describe the relationship between farming practices and climate change is that as long as farmers can continue to improve production, they will be motivated to put sustainable methods into practice, and as a result, we will be sure to see emissions reductions.

It's a subtle yet powerful distinction, and one that only recently crystalized in my own mind when I had the privilege of meeting two carbon capture entrepreneurs from West Africa at Stanford during the spring of 2023. When I asked them whether people in their countries considered climate change a top priority to address, both men answered that the effects of climate change, and not climate change itself, were what their fellow countrymen cared about most. Crop farmers, for example, have noticed more severe droughts, which could have disastrous effects on production. There is nothing political to these farmers about produce wilting on the vine. Their stakes are simple: to make ends meet by growing good crops. The only way they would think to advocate for climate policy is if it would contribute to their financial well-being. In cases like these, incentives can add an extra layer of benefit to ensure successful outcomes, with the hope that one day the technologies they support will become cost-effective enough to sustain themselves.

The story is no different here in the US when it comes to sustainable farming practices, such as no-till farming, which preserves a farmer's land and ultimately will serve to lower emissions. For centuries, the tilling and the churning of the farm topsoil has been a common practice used to aerate and prepare a field for planting. However, the trade-off has been soil erosion and nutrient runoff and the release of greenhouse gases stored in the soil. Conversely, in a no-till system, seeds are planted directly into undisturbed soil, an approach that leaves crop residues such as husks and stalks in the cultivation area to protect the soil from wind and water erosion.

Besides preventing the release of GHG from soil, no-till farming also reduces the use of fossil fuel–powered machinery by about 588 gallons of diesel—enough energy to power over 720,000 homes for a year. As a result, an estimated total of 5.8 million tons of carbon dioxide emissions are prevented by no-till farming in the US each year, which is the equivalent of taking more than one million cars off the road. In addition, no-till farming can help cut soil erosion by more than 80 percent, keeping sediments on the land and out of our water supplies.

Planting cover crops such as grains, grasses, or legumes can have many benefits, including protecting soil quality as well as carbon sequestration. As you have already learned, cover crops are not meant to be harvested, but instead offer a natural solution to issues such as soil erosion, weeds, and pests. Essentially, these cover crops do much of the same job that lost grasslands that formerly grew adjacent to farmland used to do, and are a low-cost, immediate-impact solution.

In addition to countless agricultural and environmental benefits, sustainable agricultural practices can be much more cost-effective than traditional methods, offering their own intrinsic financial incentives to farmers. Macauley Farms, which produces beef, corn, and other crops in upstate New York, for example, reported a 69 percent decrease in GHG emissions in one year. The farm also reported a 99 percent decrease in sediment losses and an increase in profit by over $25,000—a 135 percent return on investment—as a result of sustainable practices such as the ones I just mentioned.

But agricultural solutions don't just start and end on farms. In fact, a large percentage of agricultural-based emissions comes from urban and suburban communities that are miles away from growing fields. I'm talking about the enormous amounts of waste from food gone uneaten, which amounts to about one-third of all food produced globally. That's about 1.3 billion tons of fruits, vegetables, meat, dairy, seafood, and grains that either never leave the farm, get lost or spoiled during distribution, or are thrown away in hotels, grocery stores, restaurants, schools, or home kitchens. In the US alone, wasted food is worth about $408 billion per year, and uses up 18 percent of all US farmland and four trillion tons of water. Moreover, when we waste food, we also waste all of the energy it takes to grow, harvest, process, transport, and package it.

Wasted food winds up rotting in a landfill and releases vast amounts of methane. Studies estimate that if we stopped wasting food as a nation, it would be the equivalent of taking 32.6 million cars off the road. Besides emissions, of course, is the tragic paradox that much of the world, including a staggering twenty-six million adult Americans, goes hungry every day while perfectly good nourishment is being tossed out.

Not surprisingly, especially if you looked into my fraternity-house refrigerator a few years ago, the biggest proportion of food waste—about 37 percent—happens at home. Thankfully, there are plenty of actions we can take at the consumer level to make a significant difference. From shopping and cooking smarter, to freezing or sharing leftovers, to composting. The 28 percent of food waste that occurs in restaurants, school cafeterias, and grocery stores could also be curbed by practices such as limiting self-serve buffets and participating in food recovery programs and soup kitchens. As consumers, we can choose to shop at grocery stores that have committed to reducing food waste and are transparent about their progress. A whopping 8 percent of global carbon emissions come from food waste. In comparison, passenger vehicles (which hold far more of our focus) emit nearly the equivalent, at 10.8 percent. In essence, tackling food waste would nearly equate to the impact of moving to zero-emissions passenger vehicles. And while the United States definitely has work to do on this, China and India (especially with a growing population) share the bulk of the burden.

As you've seen, nature-based solutions can and should be one of the first lines of defense in any climate action plan, and deserve much more attention than they have been getting at federal, local, and state levels. Because they have the potential not only to drastically lower emissions but also to restore America's rich and diverse land, conservation efforts such as the ones I've discussed throughout the chapter are an easy win for bipartisan support.

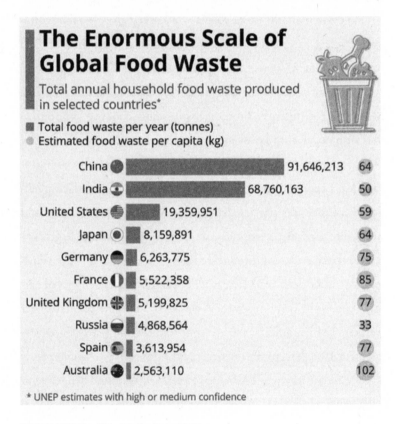

The Enormous Scale of Global Food Waste

Total annual household food waste produced in selected countries*

- ■ Total food waste per year (tonnes)
- ● Estimated food waste per capita (kg)

Country	Total food waste per year (tonnes)	Estimated food waste per capita (kg)
China	91,646,213	64
India	68,760,163	50
United States	19,359,951	59
Japan	8,159,891	64
Germany	6,263,775	75
France	5,522,358	85
United Kingdom	5,199,825	77
Russia	4,868,564	33
Spain	3,613,954	77
Australia	2,563,110	102

* UNEP estimates with high or medium confidence

SOURCE: UNEP Food Waste Index Report 2021

Nature-based solutions empower us to take immediate climate action and should take some of the pressure off clean energy to solve all of climate change before its time. Now let's take a closer look at that rough-and-tumble world of clean energy.

The 411 from Chapter 6

- The *Lorax* mindset gets it wrong when it comes to a hands-off approach to forest management. We must manage our forests to reduce wildfires and decrease carbon emissions.

- Conservation is better than preservation.

- States such as Florida and Alabama—which work with private landowners, hunters, and fishermen in conservation efforts—are often getting it right.

- Sustainable agricultural practices and the restoration of wetlands, grasslands, and marine ecosystems are affordable and immediate climate solutions.

- Decreasing food waste is imperative to solving climate change.

7

Innovating Our Way Out of the Energy Crisis

On a gray, drizzly afternoon in April 2022, I walked the grounds of Georgia's Vogtle Electric Generating Plant, where two giant gray cylinder-shaped reactor facilities have stood for over thirty years, and two new ones are underway. Although this may sound like a scene from a dystopian movie, I personally felt exhilarated to see the future of clean energy in the making. At the time of this writing, Vogtle Unit 3 is set to go online within a matter of months, becoming the first new reactor in service in the US since 2016. Once completed, all four reactors will power one million Georgia homes and businesses at once while creating zero carbon emissions. And the little waste the reactors generate will eventually be upcycled for more energy.

As I strolled the periphery of the facility, I couldn't help but notice a field of solar panels across the street that seemed oddly out of place. When I asked my guide why these panels were there, he explained that Vogtle board members wanted to diversify their energy sources as much as possible to provide customers with the best options in terms of cost and reliability. This answer made sense because, when the sun is shining, the energy from those panels would be extremely affordable. Yet on an overcast day such as the one we were experiencing, those panels had 0 percent efficacy—and only average an annual 10 percent output of their total capacity. To make up for the panels' unreliability, the company was using its supply of natural gas or coal, which, like nuclear energy, are reliable twenty-four hours a day, regardless of weather conditions.

As I listened to the explanation, I gazed out at the white, round shape of the newly built reactor facility looming in the near distance. I thought of the word *trade-offs* once again, and a new image began to form in my mind. I realized that the race toward clean energy is not a one-way track with a single finish line. Rather, it is more like an ongoing relay race in which all contenders—oil, gas, solar, wind, nuclear, hydropower, geothermal, hydrogen, LNG, and, for now, coal—participate as members of the same team, passing the ball back and forth to one another whenever it makes sense to do so.

Sticking with the relay race analogy for a moment, the renewables game gets a little more complex when you consider that we are not only passing one ball back and forth, but three, which represent the three main trade-offs we are working with in the game: affordability, reliability, and, of course, emissions numbers. Since no single solution has been able to boast all three qualities yet, we will continue to have to choose on a continual basis what source should be running, based on what trade-offs are worth it at the time.

A quick breakdown of the runners we are dealing with in the energy race will show you just how many moving parts are at play when finding the right energy combinations. Oil and coal, as we know, are both reliable and affordable, but the dirtiest of all energy choices in the race; therefore, we should try to limit their participation and let them sit out on the sidelines as often as possible. Solar and wind are clean, but only somewhat reliable, and still not very affordable for much of the world. When we need cleaner energy, and conditions are favorable (i.e., the sun is shining and the wind is blowing), we can use those as front-runners, but be ready with replacements.

Nuclear is clean and reliable, but still highly unaffordable, even though developers predict the cost will drop over the next seven to ten years with the arrival of smaller adapters and replaceable parts. Geothermal energy is affordable, reliable, and the cleanest of renewables, but subject to intensive regulation since most of its sourcing sits on federally owned land.

As the world's most abundant element, hydrogen can be extracted from fossil fuels, biomass, and water. It is versatile, because it can be used with renewables, nuclear power, and fossil fuels and applied to various sectors of the economy. Most importantly, depending on how it is processed, hydrogen can be quite clean. The Hydrogen Council estimates that using hydrogen fuel in the transportation, power generation, and heating sectors could abate up to six gigatons of CO_2 globally by 2050, a reduction equivalent to removing forty-three million cars from the road.

Making hydrogen fuel more affordable should be a top priority in the clean energy conversation. Green hydrogen, the cleanest form, currently costs between three and twenty-six dollars per kilogram. The DOE has said that in order for green hydrogen to be useful for industrial applications, that cost would need to come down to one dollar per kilogram, and that cost decrease is imperative to making renewable energy a lot cheaper. Many startups have devoted themselves to finding new ways to process hydrogen itself more efficiently, which would also bring down the price. And the IRA has offered a tax incentive of up to three dollars per kilogram for new methods of cleaner production.

Hydropower, which already accounts for one-third of the nation's renewable energy, is one of the only sources to boast all three qualities of affordability, reliability, and low emissions, and in places such as Seattle, it's a star runner most of the time. Yet it is not readily available in certain other regions of the world where water supplies are limited. These locations will have to use a combination of the other sources to keep the race moving. Liquid natural gas (LNG) is affordable and reliable, yet often still produces high amounts of methane, and therefore offers limited help when it comes to lowering emissions unless that issue is tackled.

The Texas-Size Problem with Deregulation

With all of these emergent energy source choices, it's hard for Americans to decide which trade-offs are worth it for their household or community. Many argue against the concept of energy deregulation—the concept of being *pro-choice on energy,* if you will—at all. Of course, the idea behind deregulation of energy is to increase market competition, drive up quality, and drive down costs for consumers. On its face, it puts the power back in consumers' hands and gives them greater flexibility to decide which energy packages suit their individual needs. It is especially appealing to environmentally conscious Americans who want to support green energy, even if it means paying higher prices for it than for fossil fuels.

This all sounds good; however, considering that only a handful of energy companies currently dominate the space, the competition that deregulation encourages might not be enough to make a real difference in lowering customers' costs. Also, because only large companies have the resources to catalyze new technologies such as nuclear and hydrogen power, deregulation could hinder the progress of large-scale, long-term solutions. Another factor to consider is that with privatization comes the need for controlling and monitoring distribution to ensure that customers are getting their energy needs met. This means that companies will have to invest more time and money into departments solely dedicated to this purpose.

The potential for corruption is always a risk an industry takes with deregulation. Depending on the state, energy companies most likely will find loopholes in regulations and use them to increase their profits and strengthen their market position, all at the expense of the customers. Another foreseeable problem with deregulating energy is that the few large companies in control could potentially form cartels to keep prices high through price rigging.

On most issues, Democrats traditionally lean toward regulation and Republicans support deregulation; however, when it comes to energy and all of the implications I've just mentioned, the choice is not so black and white. As you've just seen, at the end of the day, deregulation in the energy sector often favors the companies more than the customers. It also has the potential to impede America's long game for a clean energy future. The story of one particular experiment with deregulation will show you just how complicated the trade-offs can be.

Texas is a state in which most energy has been deregulated since 2002, with more than 650 different power plants owned by various companies and five major utilities responsible for delivering their electricity across 46,500 miles of transmission lines. For the five-out-of-six Texans who are free to choose their own energy retail providers, this menu of providers can be daunting and leave people feeling clueless. While some customers have saved on their bills either because they have done their due diligence or out of sheer luck, many have actually been worse off and have paid higher than the average prices paid by regulated customers in the same state.

Besides cost, the reliability of Texas's deregulated energy came into question when three coal power plants and one natural gas power plant were forced into retirement, and the state's energy reserve margin dropped from 25 percent in 2001 down to 8.6 percent by 2019. By the time the historic snowstorms hit Texas in 2021, the grid could no longer sustain the demand for energy. But the deregulated companies had little incentive or obligation to restore downed power lines or weatherize wind turbines to prevent damage, because they themselves did not own them. With one company generating power, another delivering it, and another selling it, caring for Texans' energy needs at a time when they needed it most became a shell game in which no one had any real stakes.

The media and tweeters at the time told a very different story about the cause for those major outages. For many, it was easy to blame the lack of reliability of solar and wind power. All it took were a few photos of solar panels and wind turbines covered in snow to prove to Texans why those renewables were responsible for the blackouts. Others blamed Texas's nuclear, coal, and gas, when they too became unavailable for dispatch.

The truth behind Texas's blackouts is an intricate one that doesn't fit into a TikTok sound bite. Instead, such nuanced matters get dropped from feeds and replaced by sensationalist images and stories that continue to suit viewers' shrinking attention spans. I have felt the challenging task of delivering nuanced information to the public firsthand. As a regular commentator on conservative news outlets, I am always being asked extremely intricate questions and then given thirty seconds to try to answer them adequately. It's a tricky matter, to say the least. Meanwhile, behind the scenes, I have spent countless hours with scientists, engineers, and decision-makers, reviewing mountains of data in order to hash out climate change solutions.

The first step to determining the ideal energy portfolio for America's communities is admitting that none exists. Politicians need to be willing to reconsider their positions on panacea-like solutions and lay all of the cards on the table, even the ones they may have previously discarded.

The Future of Fossil Fuels

If an open mind is a friend to innovation, then assumption is its enemy. Just ten years ago, most people didn't know exactly what hydrogen was, and many dismissed it as a viable energy solution. Now, it has the potential to become a major player in clean energy. In the same vein, the nuclear energy of our parents' generation that for decades was inextricably associated with disasters such as Chernobyl and Three Mile Island is now known to be the safest form of clean energy around. Similarly, if we take another look at fossil fuels, we may be surprised to discover what role they can still play in the future of clean energy.

As I discussed in an earlier chapter, no modern civilization has ever risen to success without a simultaneous rise in carbon emissions. That's because energy is more than just turning on lights in a home or office when it gets dark. It is the lifeblood of society. We use energy to fuel healthcare, transportation, heating, communication, food production, sanitation, and economic growth, to name a few modern-day necessities. And, if you'll recall, fossil fuels are not only used to manufacture nearly every product in our homes and businesses, but they are actually a component of many of these products, from plastics to soap to chewing gum. If fossil fuels simply disappeared overnight, as many in Washington seem to be proposing, then the world as we know it would too.

As many of us learned in physics class, the law of conservation of energy states that *energy is neither created nor destroyed; it is simply transformed from one form to another.* The great paradox of energy is that it has always existed, even before the beginning of time, yet it is always and forever changing, making it at once indestructible and ephemeral. We see energy's mutability reflected in the ups and downs of gas-pump prices and electricity bills, which depend on the ever-shifting external factors of politics, logistics, and supply and demand. The only fact we can be certain of in the energy market is that as the world continues to grow, we will always need more energy.

And as global energy demand rises steadily—it's actually doubled over the past fifty years—we need to find new sources to add to our existing supply. Notice that I said *add* and not *replace*. Because until we can figure out the right formula for energy sources that are clean, affordable, and reliable, the oil and gas we have been using remain a necessity. Even with the trillions of dollars poured into solar and wind projects over the past two decades, the world's dependence on fossil fuels has declined a mere three percentage points, from 87 to 84 percent. If you look at the chart below, you'll see that never in history has the world ceased using an energy source just because a new one came on the scene.

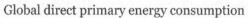

Global direct primary energy consumption
Direct primary energy consumption does not take account of inefficiencies in fossil fuel production.

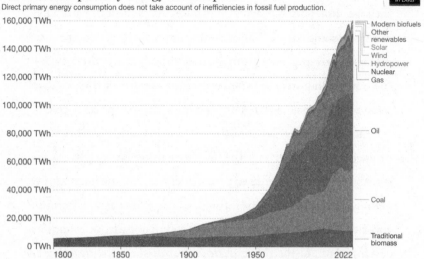

SOURCE: Energy Institute Statistical Review of World Energy (2023); Vaclav Smil (2017), ourworldindata.org/energy ·CC BY

I've already discussed in the last chapter the government's attempts to curb our reliance on foreign oil and gas by adding to the mix biofuels such as ethanol, which is derived from corn. As we know, the trade-off with biofuels is the taking up of land that could be used for food production, the conservation of important ecosystems, and carbon sequestration. Biofuels are also expensive to produce, require large amounts of water, and, although they can be slightly cleaner than fossil fuels, they actually require fossil fuels for their production. In the end, because they still give off various GHGs, the downside of such a solution is clearly not worth using it.

Another reality we must face now, before we even begin to explore a proper energy portfolio for the future, is the fact that much of the world outside of Europe and the US remains in energy poverty. The current number of people around the world who live without electricity is nearly 775 million, 85 percent of whom live in sub-Saharan Africa. These communities are forced to do whatever they can to generate power, including burning dirty fuel—and even trash—in order to survive. Clearly, until we find a way to resolve this type of abject energy poverty, the world's lofty goal of reaching net-zero emissions by 2050 remains nearly impossible.

In countries such as India, where the standards of living are rapidly increasing, the demand for energy will continue to grow with it. In earlier chapters, I discussed the recent achievement of developed nations to decouple GHG emissions from economic development. While this milestone is indeed something to celebrate, it's important to note that it has taken over a century to accomplish. In places where infrastructure and resources are vastly scarcer, who knows how long it will take before a growing economy can work together with emissions reductions, and energy sustainability.

A Changing of the Guard

Therefore, it should come as no surprise that the amount of global power generation from fossil fuels has remained relatively unchanged since 2000 and still accounts for two-thirds of the world's energy supply, with coal making up 38 percent of the world's supply, and gas about 20 percent. In absolute terms, the demand for fossil fuels has actually increased by 70 percent since that time, as it keeps up with global development.

Fossil fuels are indubitably here to stay—at least for the time being. Supplementing them with renewables and finding ways to make oil and gas cleaner are the top priorities for many scientists and engineers across the country. Recent improvements in efficiency, such as combined heat and power technology, which generates both elements from a single fuel source, can, for instance, have double the efficiency of the traditional central station power generation. Likewise, waste heat to power technology improves efficiency by capturing heat that would typically be vented and uses it to make electricity at no additional emissions cost.

Studies have shown that by equipping coal and gas plants with the latest hardware and software, their combined emissions can be reduced by 11 percent—that's the equivalent of taking 95 percent of all cars off US roads. And simply shutting the worst-performing quartile of existing coal production and keeping the rest in operation would have an enormous impact on emissions, immediately removing twenty-five metric tons of methane from the atmosphere. So far, 350 coal plants—half the total number of plants—have retired since 2010, or have plans to shut down.

Yet, rather than simply closing all of these facilities, an emerging movement toward repurposing them has come about, indicating an evolutionary approach that is much more in keeping with America's climate action goals. As I've already mentioned, transitions to clean energy are taking place in coal plants throughout the country. In Illinois alone, eleven plants have been scheduled to convert into solar farms and battery storage facilities. In Virginia, plans to segue a former coal plant into a hydrogen plant are underway. And in Massachusetts, coal plants along the coast will soon connect with offshore wind power.

The Ljungström factory in Wellsville, New York, which sits nearly three hundred miles from the Long Island coast, has for a hundred years sold parts to coal-fired plants. Yet recently it too has pivoted to offshore wind turbine manufacturing and hired 150 new workers to support its revival. It's a seismic shift for Wellsville, a mostly Republican town that for the first few decades of the last century processed up to ten thousand barrels of oil a day. But when cities began weaning their electric grids off fossil fuels and businesses around them started to collapse, decisionmakers at Ljungström decided to take advantage of their experience with steel manufacturing and develop their own line of turbines.

The San Rafael Energy Research Center, which I discussed earlier in the book, is a Utah coal plant that has joined climate action as well and is finding ways to keep their energy-producing community vibrant and relevant. Besides making coal cleaner, projects involving other forms of energy include solar, wind, and the processing of molten salt as a nuclear fuel carrier.

Because nuclear energy already accounts for 55 percent of America's carbon-free energy, the move from coal to nuclear seems like a natural evolution. The economic opportunities that coal-to-nuclear conversions could provide include between 15 and 35 percent savings in cost on nuclear construction. In addition, using existing land, grid and transmission connections, and switchyards could save millions of dollars in upfront costs. Finally, employment opportunities would be made available to the tune of 650 jobs spread across a single nuclear plant, its supply chain, and the surrounding community, with wages about 25 percent higher than those in other energy technology fields.

TerraPower, located in Bellevue, Washington, is a Bill Gates–backed nuclear innovation company aimed at lifting the world out of poverty through clean energy. One of its many projects is the repurposing of state-of-the-art utility companies for nuclear development. PacifiCorp is one of its latest collaborators, with which TerraPower plans on building a 345-megawatt Natrium™ nuclear reactor at the site of a retiring coal plant in western Wyoming. Wyoming has been a national hub for coal production, and the convergence of PacifiCorp and TerraPower would be a milestone event in the evolution of clean energy.

For our second *Electric Election Roadtrip* episode, I sat down with Dr. Joshua Walter, the principal project manager for integrated systems at TerraPower. He spoke about the massive amount of energy the world uses, and the incredible scalability that nuclear power possesses to meet this demand. We also discussed the stigma that nuclear energy has had in recent decades, and how that perception is starting to change. In 2020, for example, nuclear energy was discussed for the first time at the Democratic National Convention as a viable option for the world's future. During our interview, Dr. Walter confirmed what many already are starting to accept: among carbon-free energy sources, nuclear is the safest option out there. He also stated that for nuclear to work, long-term bipartisan support throughout changing administrations is absolutely necessary.

One place where science already is taking precedence over politics on the clean energy front is the US National Laboratories and Technology Centers, a system of seventeen cutting-edge labs overseen by the DOE. In December 2022, the Lawrence Livermore National Laboratory in California announced the first achievement of fusion ignition in history, a milestone event in which the output of energy was greater than the input. This scientific breakthrough could be a game changer in climate action because it moves the world closer to limitless, clean, safe energy.

Carbon capture, utilization, and storage (CCUS) is another technological frontier that is helping to mitigate fossil fuel impact as we transition from old to new sources, with the potential to store more than 90 percent of CO_2 emissions from power plants and industrial facilities. The promise of achieving a 14 percent reduction of global GHG emissions by 2050 makes CCUS a game changer and is why it has received substantial bipartisan support.

Since 2020, thirty-five large-scale CCUS projects have been in operation throughout the world, and about three hundred more are being developed. While the technology has been around since the 1970s, only recently has it been applied to power generation. Currently, 450,000 miles of pipelines exist in the US for transporting carbon dioxide to underground storage sites that include oil and gas reservoirs, porous rock, coal beds, basalt formations, and shale basins, all of which can hold on to carbon for centuries to come.

Yet CCUS, especially direct air capture (DAC), is still expensive and energy-intensive. The federal government has made efforts to alleviate some of the cost by setting aside $3.5 billion for DAC demonstration hubs around the country, and the DOE has also launched an initiative to drive down prices. Over the past several years, Congress has instituted many other key financial incentives related to carbon capture, including the FUTURE Act, which extends the Section 45Q federal tax credit for CCUS.

Some storage sites, including oil and gas reservoirs, also allow for enhanced economic opportunity when carbon is injected to extract additional oil in a process known as enhanced oil recovery (EOR). The extra revenue can help offset the cost of capture technology at power plants and industrial facilities.

While CCUS seems promising, operations would need to scale quite a bit to reach projected goals of emissions reductions. Right now, annual use of carbon capture is at forty-four megatons of CO_2, a number that's nothing to sneeze at. Still, to achieve net zero by 2050, we would need to capture roughly 1,300 megatons per year.

Besides carbon capture, one of the quickest ways to reduce GHG emissions in the US without eliminating fossil fuels requires no new infrastructure, mining, or importing of resources, is relatively inexpensive, and takes up no additional space. In 2022, I visited Hamm Institute for American Energy at Oklahoma State University and learned about the deployment of drones to detect methane leaks in orphaned wells, the name given to abandoned oil and gas wells that are no longer in use. Traditionally, this type of work has been done either by individual personnel on foot or by helicopter. The use of Unmanned Aerial Systems is far more cost-effective because the drones can cover thousands of miles in a short amount of time. Some of the largest companies in oil and gas are already investing heavily in drones because of these wide-ranging benefits.

Some studies estimate that 70 percent of methane emissions associated with oil and gas production can be mitigated by plugging up orphaned wells. And, based on natural gas prices over the past five years, 40 percent of those leak repairs can be achieved at no extra net cost. The Bipartisan Infrastructure Law has already allocated up to $4.7 billion in grants to help tackle the orphaned wells problem, and more big funding at a state level is underway.

Until the innovation gap in renewable energy closes, we need to find other similar ways to lower emissions of existing energy supplies. The IEA's goal of achieving a 75 percent reduction of fossil fuel emissions by 2050 will only be possible if more countries join in the Global Methane Pledge to lower impact. Emissions intensity of oil and gas operations varies greatly among countries, with the best performers emitting an amount one hundred times lower than the worst performers, among which the US has unfortunately been counted, as you can see in the chart below.

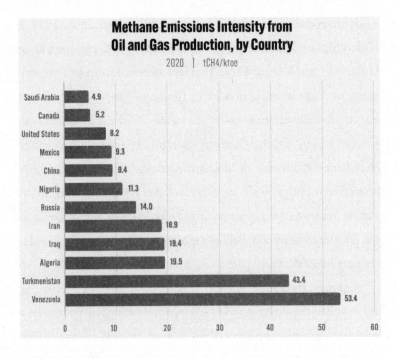

Methane Emissions Intensity from Oil and Gas Production, by Country

2020 | tCH4/ktoe

Saudi Arabia	4.9
Canada	5.2
United States	8.2
Mexico	9.3
China	9.4
Nigeria	11.3
Russia	14.0
Iran	18.9
Iraq	19.4
Algeria	19.5
Turkmenistan	43.4
Venezuela	53.4

SOURCE: Canadian Energy Centre

A Reality Check on Solar and Wind

As we all know, employing the wrong action for the right reasons still does not make the action right. When we look at the facts, relying solely on intermittent renewables such as solar and wind power as main energy sources makes no sense. When these energy sources first came onto the climate action scene, they seemed to be the best bets to lower emissions in a short amount of time. Countries invested heavily in panels and turbines, pledging to stop using fossil fuels altogether within a matter of years. Yet, as we have seen with the case of Germany, relying on such unreliable energy can lead to disastrous shortages, financial hardship, and dependence on foreign governments.

Besides being literally as undependable as the weather, solar panels are also still cost-prohibitive for most mid- to low-income households. Even after the current federal solar tax credit, the average cost to buy and install rooftop panels in the US is about $20,000. Only about a quarter of these costs are panel-related; the rest are for labor, operational expenses, and additional equipment, such as inverters and control circuitry. In addition to upfront costs, solar panel users also need to plan for inevitable repair and maintenance as well as higher electricity bills.

As of now, households where governments mandate solar power pay 11 percent more in monthly utility bills than those in other states. In California, residents pay a whopping 80 percent more for energy than the US average. Likewise, in Germany, a country that has invested hundreds of billions of euros into building massive solar and wind farms, household energy bills increased 34 percent between 2010 and 2020. And as we all know, Germany has been facing a major energy shortage that has forced the country to rely on Russian natural gas to supplement its intermittent renewables.

In addition to reliability and cost, the question of just how clean solar energy really is needs to be considered. I have already discussed the enormous toll on the Earth that results from the mining that solar panels require: quartz, silver, aluminum, and other minerals. Add to this the diesel-powered machinery used to make these panels and the fossil fuels used to transport them, and we are looking at three years of solar panel usage before a solar panel simply pays off its carbon deficit.

In addition, these panels have a short lifespan, and it is predicted that by 2050, the waste from used panels will amount to seventy-eight million tons, the equivalent of throwing away sixty million Honda Civics. The infrastructure needed to create a circular economy around such vast amounts of recyclable waste will require enormous investments. Similarly, breaking down the technology of EV batteries to reuse its metals—90 percent of which are indeed reusable—is a daunting enterprise that must be calculated into their overall value.

Another factor to take into consideration with solar and wind power is the physical space each requires that could be used for other purposes. One recent study done by The Nature Conservancy assessed the long-term clean energy needs of eleven western states, including charging stations for tens of millions of electric cars. (This study, of course, assumed that intermittent renewables and EVs played a major role in these states' energy future.) The good news is that the study took into account that certain areas, such as national parks, wildlife refuges, wetlands, and endangered species' habitats, would be off-limits. The not-so-good news is that the study concluded that roughly twenty-one million acres—that's a little less than half the size of Utah!—would be needed to accommodate solar and wind farms and charging stations.

Similarly, to paint a clearer picture, there are 12 nuclear plants scheduled to close by 2030. With today's technology, it would take roughly 5,172 wind turbines (covering over 337 million square feet of land) or 37.5 million solar panels (covering over 500 million square feet of land) to replace that power. *If solar and wind aren't possible, it would take 19 new coal plants to replace the 12 nuclear plants, spewing an additional 81 million tons of* CO_2 *into the atmosphere annually.*

Historically, local communities across the country have resisted solar and wind because of the visual obstructions and noise pollution they bring with them. Farmers also oppose the usage of land that could go to food production. If current decision-makers did have their way in developing solar and wind to the degree they want to, and if those technologies did in fact grow in reliability and longevity, would we really want to live among hills and plains dotted with turbines and panels? Talk about a dystopian landscape!

In early 2023, photos circulated of German police officers carrying away Greta Thunberg by her arms and legs after she sat in protest in front of a coal plant that was scheduled to expand. The irony of this scene is striking. The solar/wind-only mindset the young Swedish activist is known for is the very same sentiment that required that coal plant to reopen in the first place.

It's time for environmental activists and politicians to acknowledge the drawbacks of certain proposed solutions so they can start looking at better ones. When I first began work with ACC, I saw clean energy through a black-and-white lens, much the way Greta does. Renewables were good, fossil fuels were bad, until I considered the trade-offs of each. US drilling for oil was terrible, until I learned that if we didn't do it here, we would have to outsource it from countries with much dirtier and ethically questionable operations.

Pushing a solar and wind agenda in the hopes that these sources of energy will become more affordable, reliable, ethical, and even climate-friendly is one of the greatest travesties of the climate action era. Just like the rest of us, the drafters of the GND cannot guarantee outcomes. Their blanket approach to clean energy has been fabricated on too many unknowns, and now that blanket is quickly becoming threadbare.

Maybe we will be able to find ways to store solar and wind power to use on days when the sun isn't shining and the wind isn't blowing. Maybe we will be able to bring down costs of panels and turbines for lower-income households to afford. Maybe we will become less dependent on China and be able to manufacture solar panels and turbines on domestic soil. That's a whole lotta *maybes*.

Still, in June of 2022, the US federal government doubled

down on intermittent renewables and announced a two-year pause on imposing any new solar energy tariffs. While this move sounds to clean energy enthusiasts like a step in the right direction, it focuses us on the wrong solution. Furthermore, focusing on solar and wind power detracts from investments that should be made in other solutions, including the ones I've discussed in this chapter, that have far fewer inherent problems. At its core, we need to solve the gap below between the "current policies" and the 2 degree and/or 1.5 degree pathways. As we've seen throughout this chapter, we can't solve that gap without continuing a great deal of innovation. That's up to our generation to solve—and it's actually an exciting challenge.

My hope is that by now you have a better understanding of just how nuanced the conversation about clean energy is, and how much of its success hinges on factors beyond colliding neutrons, cobalt extraction, and photovoltaic cells. Our nation and our planet need us to pay attention to the nuanced truths behind the sound bites and political branding our generation is used to and chase after innovative solutions we haven't even dreamed up yet.

As students, entrepreneurs, voters, and problem solvers, we owe it to ourselves and our planet to engage in a rational climate change conversation that accounts for trade-offs. It's our choice to act reasonably and responsibly based on facts that are far from black and white. In the next and final chapter, I'd like to explore what this type of discernment will look like when we put it into action.

The 411 from Chapter 7

- We need to close the innovation gap with more incentives and support for clean energy at local and state levels.

- It's time to redefine the renewables race to include a wide portfolio, with nuclear and hydrogen at the forefront. Solar and wind definitely can't do it alone.

- Phasing out fossil fuels overnight is a failed strategy—and it will continue to fail. With growing demand for fossil fuels, we must find a balance between decreasing our dependence on them *and* improving their efficiency.

- Repurposing state-of-the-art utility facilities for nuclear and other renewables is a key step in transitioning away from fossil fuels.

A True Climate Commitment

When I walked into the 2019 Congressional Black Caucus in DC, I was surprised to discover that the event's flagship sponsors were oil and gas companies. After growing up hearing endless criticism of Republicans who were *bought out* by oil and gas, I was stunned to see a room full of Democrats schmoozing with the so-called enemy. A few minutes later, Representative John Lewis (D., GA) saw my slightly bewildered expression and explained to me, "Benji, we might not like it, but we have to work with them because that's just how the world works." I'll never forget those wise words told by a man who knew better than anyone else how to put aside differences to achieve a greater goal.

As one of the Big Six of the 1960s Civil Rights Movement, Congressman John Lewis achieved what many in his day deemed impossible. He did this not by inciting fear or condemnation, but simply by pushing consistently against the status quo, a gentle but bold approach that required patience, commitment, and courage. It also required an intelligent diplomacy, a willingness to cross clearly drawn lines, and a mustering of mutual respect on both sides to see progress made.

Officials who find themselves working with the old guard of fossil fuel giants are not *hamstrung*, as some like to call it, or *bought out*. They are simply operating within the bounds of reality. Acknowledging how far the fossil fuel industry has carried us as a society, they are committed to finding ways to work with these companies, not against them. In the same way, Democrats and Republicans might not see eye to eye on several issues, but we can still find common ground in reaching our shared goal of fighting climate change.

When I started American Conservation Coalition, only two Republicans in Congress were talking about climate change. Now there are almost a hundred in the House of Representatives alone who are a part of a caucus—the second largest one in Congress— dedicated solely to climate action. As more and more conservatives join the conversation, I see change happening that I never would have dreamed of. Much of this progress has started with something as simple as a conversation.

In the fall of 2022, I texted a few members of Congress I know personally to invite them to the home of Democratic US Representative Dean Phillips, who graciously agreed to play host to a dinner for all of us. Even though during the 2018 elections, ACC endorsed his opponent, Dean and I have since become friends, partly because of our shared desire to promote bipartisanship to solve important issues. This dinner would be our first concerted effort toward that goal, with three Democrats and three Republicans on the guest list just months before the new congressional session, when political posturing was reaching a climax.

That evening, as we first greeted one another, the nervousness in the room was palpable. I could guess that no one present had ever done anything like this before. I myself had no idea what to expect from these other men as we sat down to share a meal in such an intimate setting—gloves off, no agenda, away from the public eye. (And no wine to boot!)

As we sat down, I quickly saw that I had nothing to worry about. Without anyone airing so much as a single grievance, conversations began to flow freely among the guests, with topics ranging from clean energy innovation to the role of government in project management, to the importance of national parks and public lands. When we started getting into the weeds about specific policies, I was sure that the two moderators I'd invited to the dinner would need to kick into high gear. But I was wrong about that too.

Judging from the looks of surprised delight on all the guests' faces over the course of the evening, they were just as taken aback as I was by how much overlap and common ground the well-intentioned individuals from both sides shared on nearly every issue. Some of them admitted that they had never even spoken to one another until that night, even though they had had ample opportunity to do so throughout the years. I thought, *If our politicians aren't even talking, then how in the world are we expected to accomplish anything meaningful?*

Even more compelling were the reactions of top leaders afterward when I told them about the dinner. "How in the world did you manage to pull that off, Benji?" they asked. I didn't quite know how to answer them. The question itself said a lot. Clearly, no one else had tried to do something like this before, even though it was as simple as a twenty-four-year-old guy who loves nature and cares about the future of our planet sending out a few friendly texts.

As you've seen, historically, conservatives have led the way on environmental action, which was never seen as a partisan issue until the early 2000s. Even as late as 2008, in an effort to boost bipartisan support, Democratic Speaker of the House Nancy Pelosi and former Republican Speaker Newt Gingrich did a national television ad together, calling upon Americans to put aside their differences to fight climate change. That was just fifteen years ago, yet it feels like light-years away from the politically polarized world we live in today. Still, according to the polls and countless conversations with my peers, our generation does not see climate action as a partisan issue at all. More than any other motive, our sincere concern for the natural world is what propels so many of us to take action.

Continuing in the footsteps of conservative environmentalists in the last century, a new set of common-sense climate solutions is in order. The Green New Deal, with its lopsided methodology and calls for sweeping change, is simply not the answer. Nor is the endless criticism and fear-mongering that flood our social media feeds for both sides. As you have seen throughout these chapters, a better, more holistic approach exists that includes protecting our economy, our country's natural resources, and the daily needs of America's various communities.

It's time to restore balance by bringing conservative voices back to the conversation. Just as we need to look at both sides of the carbon equation—that is, its output and its sequestration—we must also take an intelligent approach to climate action that balances benefits with trade-offs. With each solution, we must consider its initial financial cost as well as its long-term economic viability, reliability, and scalability. We must also assess the solution's actual climate impact in terms of emissions versus the toll it will take on the physical environment. Equally important is our ability to analyze strategy through the lens of short-, medium-, and long-term results as well as scalability on a global level. Only then will we be able to discern what every community, state, and country in the equation needs to do to contribute meaningfully to the fight against climate change.

As young people who care passionately about our planet, it's up to us to lead the way back to bipartisan climate action. Leading the way means respecting the generations of oil and gas workers who until now have powered our homes and businesses, and helping them pivot to cleaner solutions. It means including the needs of the millions of rural Americans whose hard work makes this country thrive and creating smart incentives that support their work. It means restoring our beautiful forests, oceans, wetlands, and grasslands, taking care of them so that they can continue to take care of us. It means encouraging the spirit of innovation and entrepreneurship that will make us energy independent and help us have a powerful influence on the rest of the world.

At an event with leaders from Vatican City and Stanford Graduate School of Business, Dr. Arun Majumdar, the inaugural dean of the Stanford Doerr School of Sustainability, spoke about the power of failure. He said that one way you know that something is about to become big on the marketplace is by how many other companies around it have failed or are failing. Think about how many companies have surfed the tidal wave of social media over the past three decades—Myspace, ASKfm, Google+, and countless others—but are no longer around today. If the wave hadn't been so enormous, it wouldn't have attracted these businesses to begin with.

Similarly, the evolution of clean energy has been taking place for some time, and the private sector will play an even greater role in its innovation. Therefore, we will need substantial critical mass to find those entrepreneurs whose ideas will withstand the resisting tides, the Elon Musks who will push back against the status quo. Whether you're interested in developing a new mode of transportation, or restoring public lands, or cleaning up oil and gas, a little Lucky Girl Syndrome can go a long way. Government funding already exists to support your new ideas, and think tanks such as Greentown Labs are out there welcoming entrepreneurs and innovators every day. When we first started ACC, I didn't have all of the answers—and the truth is, I still don't—but I did have an open mind and a can-do attitude that helped me go out on a limb. It's time for the entire country to do the same.

Getting Committed

As Congressman John Lewis pointed out, there's no other way to solve the dilemma in which we find ourselves than to work with what's already here. You know by now that no single set of solutions exists to reverse the damage human progress has done to the planet. And no one plan—not even this book—has all of the answers. ACC's American Climate Commitment is the set of six principles that embrace the ideas I've presented throughout this book. It is not a prescription. Nor is it a manifesto. Instead, I like to think of it as a springboard, providing young people with the vision and platform we need to engage productively in a climate change conversation—including both liberals *and* conservatives.

As we all know, climate action starts with the individual. At my current home in Arizona, I make sure to charge my phone and laptop during the peak hours of sunlight so that I don't use up valuable stored energy that can be used at night. Carpooling, combining errands to make fewer trips, shopping for sustainable products, and not wasting food are other ways we can be personally more efficient.

Making more efficient choices on a daily basis doesn't necessarily need to go by the name *climate action,* even though it does have the same outcome. My grandma would call it just being smart. Florida governor Ron DeSantis and Georgia governor Brian Kemp are known for their excellent environmental track records, yet neither man calls what he does *climate action.* The language they use, instead, speaks to what is important to the people they lead, such as clean energy, clean air, national security, a better return on farming practices, upgraded infrastructure, a robust economy, and technological innovation.

I recently met a college student from Stetson University who has a tough time talking with her conservative parents about climate action in a way that resonates with them, even though she knows they care deeply about the environment. She had even recently switched to liberal politics because she couldn't find a place where she felt comfortable engaging with conservatives her age about the environment. When she saw me speaking at the 2023 Aspen Ideas: Climate event in Miami, she was thankful to hear the language I was using around common-sense solutions because she knew her parents would be able to get behind ideas such as economic opportunity in renewables, support for rural communities, and conserving our natural habitats, all of which will ultimately help our country and our planet thrive.

Getting Consistent

Sometimes protecting the environment means doing what's hard. In 2020, I was sitting at dinner with my parents who had come to visit me in Washington state. Our outdoor table at Ray's Cafe faced the beautiful Pacific Ocean below, and the sun was just about to set. We were all smiles as we watched seals lumbering about on the rocky shore and bobbing up and down in the waves. A few moments later, we saw a fishing boat heading out onto the water with two men, one of whom held a heavy club in his hand. My parents looked surprised, then bewildered when they noticed a seal alone on one end of the shore toward which these men were heading. I, on the other hand, understood the situation in its entirety, though that didn't make it any easier to explain.

About six months earlier, ACC had collected thousands of signatures from college students in the area to help Congress pass a bipartisan bill sanctioning the controlled hunting of seals in the Pacific Northwest. The bill was part of a long-term effort to preserve the salmon population that in recent years has reached record lows due to the overpopulation of seals, their main predator. The problem started, as many disruptions in the food-chain do, at the top, with the recent overhunting of the seals' own primary predator, the orca whale.

Now, to restore balance and repair some of the damage we humans have done, conservationists have been working hard to get the count of seals back to a proportionally normal number, and have already seen improvements in salmon populations. Still, the positive outcome doesn't make it any easier to sanction the killing of an animal everyone finds irresistibly adorable. Similarly, tree thinning and prescribed burns might be a necessary protocol to preserve our nation's woodlands, yet these practices still sometimes cause us to wince inside.

By the same token, being able to do something does not mean that we should, a point I already made in the last chapter when discussing solar and wind power. To reiterate, there's nothing wrong with pursuing those intermittent renewables; in fact, they are an undeniably important part of our clean energy future. However, until we experience a breakthrough that improves their energy storage capacity, we need to continue to look for better solutions.

Preparing the national grid system for cleaner energy and developing cost-effective carbon technology and reliable energy storage are all solutions that require time and call for a patient urgency that might be foreign to many activists. Clean energy transportation modes at mass scale will not appear overnight, nor will important emerging technologies such as hydropower and geothermal fuel, which, in the long run, could be more efficient than EVs in powering our cars. Moving manufacturing of solar panels and wind turbines back home involves the slow process of weaning ourselves from Chinese production and processing. And restoring vegetation to earlier levels in our planet's history is a viable possibility that, again, will take time.

In the heat of the summer of 2021, ACC held its first-ever American Conservative Climate Rally, in Miami, Florida. Many members of Congress showed up, as well as mayors and conservative leaders from Miami and other Florida cities. Altogether, about three hundred people were present, a disappointing number for me, because I'd had higher expectations for the turnout. Yet the gathering was large enough to cause a stir among Far Left and Far Right residents who resented our presence on their turf, most likely because no one had ever held a conservative environmental rally before. They showed up with signs and heckled us throughout the entire event. Still, no anchors or reporters deemed the gathering newsworthy, and the lack of coverage was also disappointing, though not surprising.

When conservative leaders get a chance to sit down and talk about climate change, they are often met with skepticism and hostility. Frequently in interviews, I receive a slew of *gotcha!* questions, aimed at exposing some malevolent agenda I don't have. Left-leaning news outlets assume that conservatives don't really care about the environment. They ignore Republican climate leaders such as congresspeople John Curtis, Bruce Westerman, and Cathy McMorris Rodgers, who are all making an enormous difference on a large scale. Bipartisan progress is being made right now, with trillions of public and private dollars being invested into climate action, and bills being passed left and right to help mitigate emissions.

At the ACC Summit that took place in DC last year, our speakers included a Democrat member of Congress, a liberal-leaning actress, a liberal-leaning sports player, multiple Republican members of Congress, a Republican governor, and a few Republican news personalities. Twenty percent of the crowd was left of center, 15 or so were in the middle, and 65 percent were right of center. Even though ACC is the only organization out there putting together such politically diverse climate action events, there was still no media coverage of this groundbreaking event. Still, whether we make the five o'clock news or Twitter feeds or not, we are unmistakably making history. In just six years, ACC's movement has worked tirelessly to put into practice the approach spelled out in this book. Using The Climate Commitment as our guide, we have passed numerous bipartisan environmental policies for the first time in decades, helped launch the fourth-largest caucus in Congress (the Conservative Climate Caucus), and completely shifted the national narrative from conservative climate denial to cross-partisan action. We've gotten a lot done. But our work is far from over.

Getting Flexible

While steadfastness is key, it's also important to remember that all long-term ideas are purely speculative. They require us to be hopeful enough to engage in them wholeheartedly, but humble enough to know when to change course. As Ralph Waldo Emerson famously wrote, "A foolish consistency is the hobgoblin of little minds." It's time for our leaders to unhitch their wagons from yesterday's great ideas and admit that some of them have turned out to be falling stars.

Just take a look at some of our earnest, though failed, attempts at reducing emissions and our dependence on fossil fuels in recent years. Twenty years ago, ethanol held the promise of a viable clean energy solution. However, the overproduction of corn stripped our land of its natural carbon sequestration properties and contaminated our water supplies, and the production costs proved too expensive without the help of government subsidies.

Likewise, the aggressive planting of more trees to counteract deforestation initially made sense, but inadvertently transformed our nation's forests into tinderboxes that could be easily set ablaze and actually increased GHG emissions. Strict regulations and permitting processes that were meant to ensure safety have ended up hindering important domestic innovation and forced us to rely on foreign governments for clean energy technologies to the point of risking national security.

Hindsight is twenty-twenty. Yet urgent situations require us to make choices based on the imperfect data available to us at the moment. We make the most out of the information that's available at the time and do well to hold loosely to our theories. Only later can we determine whether to continue on our original course of action or try a new one.

Being willing to be wrong is often easier said than done. From politicians, it requires a certain level of humility and open-mindedness and the courage to lose popularity points at times. From the rest of us, it takes a determination to step outside of our echo chambers and consider evidence that might challenge our previous views.

Getting Loud

Former congressman Ryan Costello, who served as a US representative of Pennsylvania from 2015 to 2019, was one of the first Republicans in recent years to have a pro stance on climate change and also one of the first congressmen I ever had a chance to officially meet with. I was all of sixteen years old and nervous as hell during our meeting. But Ryan quickly put me at ease and told me how grateful he was that I'd reached out.

"You know, Benji," he said, "my colleagues and I would consider voting differently on plenty of policies if we just heard from a couple of voters from time to time." Considering that this comment came from a federal legislator, it's easy to imagine how much less a local representative might hear from their own voters.

Even at the federal level, officials hardly ever receive constructive input from their constituents, much less requests for internships or volunteer work between campaigns. Instead, they simply default to listening to the loud voices of companies that sponsor or support them. Yes, there is a lot of money in politics, but in the end, votes always matter more than money—don't let anyone tell you differently. When I was fourteen years old, I joined a team of volunteers to help state assemblyman Dave Murphy win the primary by just 224 votes. Had I and the other volunteers put in thirty instead of forty hours a week on that campaign, we may not have gathered as much support, and the outcome could have easily gone the other way.

Beyond elections, you as an individual have more power than you might be aware of to influence elected officials. You may feel your small municipality or state may not make a big difference in the larger climate change conversation. It does. Imagine if everyone reading this book started to speak to their local assembly members, representatives, senators, mayors, and governors about just one of the issues that are important to them. Collectively, those small changes would add up to a lot, enough to start to move the needle at the national level.

Exposing elected officials to new climate action ideas and projects is best done at the local and state levels first, because those are the people who will understand the problems best and be able to have the greatest impact. Congressman Dan Newhouse, a US representative for Washington state and chairman of the Congressional Western Caucus, is known as one of the most accessible congressmen in office and also one of the biggest state-level supporters of clean energy innovation. Besides steering sizable DOE funding into nuclear development collaborations that have provided hundreds of new jobs, he also has been instrumental in improving permitting for hydroelectric projects that also protect Tribal interests.

Back in 2021, Congressman Newhouse accompanied me on a tour of the Pacific Northwest National Laboratory, a world-class facility that is leading the way in renewable energy production. That day, we had the chance to see some incredible technological advances, including the largest wind and solar storage battery and the smallest nuclear reactor to date. Our guide also showed us the lab's expansive data system that monitors where energy shortages and surpluses are located throughout the grid and suggests more efficient ways to redirect those surpluses.

All of this clean energy innovation is familiar territory to Congressman Newhouse, who is one of many elected officials who are avid supporters of clean energy solutions. Many others, even though they are eager to support issues that matter to our generation, have never toured a National Laboratory, simply because no one has suggested it. Unless our generation speaks up and asks that they prioritize these issues, these elected officials will remain in the dark about what matters to us and how to help.

You don't need to know everything about a topic to set up an appointment with your local city or county councilperson to discuss it, but you have to know enough to be taken seriously. And it's not enough to just mention what you care about. It's also important that you help them to understand the complex realities behind the challenges.

Besides meeting in person, another way to expose elected officials to real-life solutions is through social media. All politicians need to keep their fingers on the pulse of their constituents, and to help, they each have a social media team devoted solely to keeping up with what supporters and detractors are saying. I promise you, if you're writing it, they're reading it.

My work with ACC has proven that we can indeed solve climate change together. Countless left-of-center organizations, universities, and television syndicates have embraced me as a conservative voice of reason and compromise. At the same time, more and more right-of-center leaders are leaning into the conversation and advocating a common-sense approach to climate action. Conservatives have the power to take swift and meaningful steps toward conservation, develop smart, limited-government policies, and make sustainability profitable.

This book and The Climate Commitment are just the start of a new, common-sense approach to climate action, one that aligns with our generation's values and makes sense economically and geographically. A balanced plan of action is in order that takes into consideration all of the trade-offs involved in each solution as well as short-, medium-, and long-term outcomes. We can move forward only if we put political tribalism aside and rise above the noise of scare tactics and denialism. Only then will our generation of environmentalists be able to stand together and fight climate change, regardless of our political ties. Our planet deserves this. So do we.

ACKNOWLEDGMENTS

THIS BOOK IS a culmination of my life experiences—applied to something I care deeply about. The manuscript (and my life in general) would not be possible without *thousands* of magnificent people who have helped me along the way.

I first want to thank Salwa Emerson, my writer, for helping me formulate ideas and powerfully deliver them. Without your eloquence and empathy, this book would not have been possible. You're the top of your craft—and also a great friend.

Next, I want to thank the RPrime team—Jonathan Roberts, Stefanie Robinson, Peter Springer, and Garr Larson. Without you, this book would be far less powerful. From our retreats to diving deep into the brand-new data presented in this book and the entire editing process . . . you are the smartest and most self-less partners I could have ever asked for. I know this is just the beginning of our work together!

ACC's common sense and "eco-right" partners have provided unparalleled guidance, leadership, and allyship in a space that didn't even exist a few years ago. A huge thank you to the staffs of PERC, ClearPath, DEPLOY, RepublicEn, CRES, R Street Institute, Bipartisan Policy Center, C3 Solutions, Audubon, Breakthrough Institute, National Wildlife Federation, and so many other organizations for all you do.

I owe a massive amount of gratitude to the entire ACC team, board members, donors, and volunteers. Fighting together for the same movement, their steadfast dedication to the cause has driven every success we've had. It's true when people say it "takes a village." I'm blessed to be surrounded by a "village" of some of the most amazing people on the planet. A special shoutout to Danielle Butcher Franz, Chris Barnard, Stephen Perkins, Michael Esposito, Quill Robinson, Matt Mailloux, Karly Matthews, Lucero Cantu, Jackson Blackwell, Alex Joyce, Madison Link Rees, and countless other team members who've dedicated their entire lives to ensuring our movement succeeds. You will forever be heroes in my eyes.

Over the past seven years of doing this work, my mentors have become some of my closest friends. I've learned how to manage an organization (and my voice) by the seat of my pants. Without Bill Bryant, Jay Faison, Matt McIlwain, Mike Vaska, Tim Higgins, Gary Rappeport, Collin O'Mara, Carlos Curbelo, Jim Connaughton, Todd Myers, Mike Wascom, Rob McKenna, Jen Lewandowski, Mrs. Karen Reeck, Bev and Tom Hanson, Dural Morris, Ana Mari Cauce, Doug Ducey, Evan Baehr, Chris Larsen, Jason Buechel, Skip Rowley, Don Bennett, Rich Powell, Ross Perot Jr., Paul Bodnar, Tony Kreindler, the Honorable Dan Evans, Chris Bayley, the late Tom Alberg, the late and Honorable Slade

Gorton, and *literally* endless other mentors, I wouldn't be anywhere near where I am today.

I want to thank my closest friends, old and new. While some friendships come and go, each of them has shaped me in a major way. They've kept me grounded—and helped ensure I keep my spunky personality amidst the seriousness of my work! While we've grown further apart in recent years, Brandon Books, Jada Taylor, Mike Bray, Andrew Syring, the Shaw family, Haley France, the Bowra family, and the late Tony Daharsh—thank you for the incredible role you played in my life. We've truly been through hell and back together. To my love, Elsah Boak, and closest friends, Johnny Ochsner, Chandler Crane, Nick Orlov, Brennan Holmes, and Nate Nagel . . . thank you for being the constant support system throughout the past few years. It hasn't been easy, and I haven't been the friend I've wanted to be, but you've been by my side through it all.

SOURCES USED

Introduction | Evolution, Not Revolution

viii **The nine years from 2013 to 2021:** Robert Rohde, "Global Temperature Report for 2022," Berkeley Earth, January 12, 2023, berkeleyearth.org/global-temperature-report-for-2022.

ix **82 percent of 18- to 35-year-olds polled:** American Conservation Coalition, Cli acc.eco/acc-youthpoll.

x **President Nixon created the Environmental Protection Agency:** Lily Rothman, "Here's Why the Environmental Protection Agency Was Created," *Time*, March 22, 2017, time.com/4696104 /environmental-protection-agency-1970-history.

xii **The governor is famous for declaring:** Deborah Gordon and Daniel Sperling, "Governor Arnold Schwarzenegger: Environmentalist," Oxford University Press Blog, January 2009, blog.oup.com/2009/01/schwarzenegger.

Chapter One | Overcoming Our Political Divide

4 **We've heard outspoken individuals:** Graig Graziosi, "Tucker Carlson Says Climate Change Is a Liberal Invention 'Like Racism' in Shocking On-Air Rant," *The Independent*, September 13, 2020, independent.co.uk/news/world/americas/tucker-carlson-climate -change-fox-news-california-wildfires-racism-liberal-b434261.html.

4 **Just as ridiculous was New York senator Charles Schumer**: Zack Budryk, "Schumer Calls for Action on Climate after Ida Flooding," *The Hill*, September 2, 2021, thehill.com/policy/energy -environment/570570-schumer-calls-for-action-on-climate-after -ida-flooding.

4 **climate activists blaming Texas's 2021 snowstorm**: Oliver Milman, "Heating Arctic May Be to Blame for Snowstorms in Texas, Scientists Argue," *The Guardian*, February 17, 2021, theguardian.com/science/2021/feb/17/arctic-heating-winter -storms-climate-change.

6 **About 28 percent of rural Wisconsinites**: Staff, "Government Officials Visit Deployment Site as Wisconsin Continues to Extend Rural Broadband Access," Spectrum News 1, August 3, 2023, spectrumnews1.com/wi/milwaukee/news/2023/08/03/wisconsin -continues-to-extend-rural-broadband-access.

6 **California's aggressive plan to ban**: Dan Avery, "These 9 States Are Banning the Sale of Gas-Powered Cars," CNET, September 7, 2023, cnet.com/roadshow/news/states-banning-new-gas-powered -cars.

8 **Trump's promise to bolster goods-producing industries**: Danielle Kurtzleben, "Rural Voters Played a Big Part in Helping Trump Defeat Clinton," NPR, November 14, 2016, npr.org/2016 /11/14/501737150/rural-voters-played-a-big-part-in-helping-trump -defeat-clinton.

8 **Start learning code:** Alexandra Kelley, "Biden Tells Coal Miners to Learn to Code," *The Hill*, thehill.com/changing-america /enrichment/education/476391-biden-tells-coal-miners-to-learn-to -code.

10 **Introduced to Congress by Representative Alexandria Ocasio- Cortez:** "The Green New Deal," Bernie Sanders, berniesanders .com/issues/green-new-deal.

12 **A simple analysis using Project Drawdown's:** "Table of Solutions," Project Drawdown, drawdown.org/solutions/table-of -solutions.

12 **The United States contributes:** statista.com/statistics/271748 /the-largest-emitters-of-co2-in-the-world.

12 **This is all for the cost of $93 trillion**: Ari Natter, "Alexandria Ocasio-Cortez's Green New Deal Could Cost $93 Trillion, Group Says," *Bloomberg*, February 19, 2019, bloomberg.com/news/articles /2019-02-25/group-sees-ocasio-cortez-s-green-new-deal-costing-93 -trillion.

13 **it costs between $5 and $100 per ton**: Carly A. Phillips et al., "Escalating Carbon Emissions from North American Boreal Forest Wildfires and the Climate Mitigation Potential of Fire Management," *Sci. Adv.* 8, no. 17 April 27, 2022, science.org/doi /10.1126/sciadv.abl7161.

13 **Even House Speaker Nancy Pelosi**: Chris Cillizza, "Nancy Pelosi Just Threw Some Serious Shade at Alexandria Ocasio-Cortez's 'Green New Deal,'" CNN, February 8, 2019, cnn.com/2019/02/07 /politics/pelosi-alexandria-ocasio-cortez-green-new-deal/index.html.

14 **former energy secretary Ernest Moniz**: Jeff Brady, "Despite Few Details and Much Doubt, the Green New Deal Generates Enthusiasm," NPR, February 8, 2019, npr.org/2019/02/08 /692508990/despite-few-details-and-much-doubt-the-green-new -deal-generates-enthusiasm.

15 **With a whopping 40 percent of Americans**: David Nadelle, "Percentage of Americans Unable to Cover a $400 Emergency Expense Shoots Back Up to Pre-Pandemic Levels," Yahoo Finance, June 1, 2023, finance.yahoo.com/news/percentage-americans -unable-cover-400-133058811.html#:~:text=According%20to %20the%20Fed's%202022,and%20back%20to%202019%20levels.

18 **The Clean Water Act of 1972 was just**: "Five Clean Water Act Success Stories," PBS, February 24, 2023, pbs.org/wnet/peril-and -promise/2023/02/five-clean-water-act-success-stories/#:~:text =More%20than%20fifty%20years%20later,goals%20has %20doubled%20since%201972.

Chapter Two | Streamlining the Complicated Role of Government in the New Green Economy

25 **overconsumption of coffee has led**: Shreya Dasgupta, "Coffee in Trouble: 60% of Wild Coffee Species Threatened with Extinction," *Mongabay*, January 17, 2019, news.mongabay.com/2019/01

/coffee-in-trouble-60-of-wild-coffee-species-threatened-with
-extinction/#:~:text=Of%20the%20124%20species%20
of,world%2C%20is%20in%20particular%20trouble.

27 **In response to these dire:** Beth Gardiner, "The Clean Air Act
Saved Millions of Lives and Trillions of Dollars," *National
Geographic*, December 29, 2020, nationalgeographic.com
/environment/article/clean-air-act-saved-millions-of-lives-trillions
-of-dollars.

28 **Thanks to the act's stringent:** "Clean Air Act Requirements and
History," US Environmental Protection Agency, epa.gov/clean-air
-act-overview/clean-air-act-requirements-and-history.

29 **Among the earliest achievements:** Beth Gardiner, "The Clean Air
Act Saved Millions of Lives and Trillions of Dollars," *National
Geographic*, December 29, 2020, nationalgeographic.com
/environment/article/clean-air-act-saved-millions-of-lives-trillions
-of-dollars.

30 **100 percent phase-out mark for chlorofluorocarbons:** "Happy
Birthday, Montreal Protocol on Substances That Deplete the
Ozone Layer!" Council for the Advancement of Science Writing,
September 16, 2022, council.science/current/blog/happy-birthday
-montreal-protocol-ozone.

34 **NEPA's review process takes:** Jenifer Perez, "The Potential
Revision of NEPA Could Lead to Millions of Dollars Saved on
Federal Projects," *Policy Interns*, March 9, 2023, policyinterns.com
/2023/03/09/the-potential-revision-of-nepa-could-lead-to-millions
-of-dollars-saved-on-federal-projects.

35 **reducing its timeline from seven years:** Perez, "The Potential
Revision of NEPA."

39 **Under Section 404:** "How Does Permitting for Clean Energy
Infrastructure Work?" Brookings, September 28, 2022, brookings
.edu/research/how-does-permitting-for-clean-energy-infrastructure
-work.

39 **While countries like China:** Adrijana Buljan, "China Now
Has 31+ GW of Offshore Wind Installed, Country on Track to
Hit Wind and Solar Targets Five Years Early, Report Says,"
offshoreWIND.biz, June 30, 2023, offshorewind.biz/2023/06/30

/china-now-has-31-gw-of-offshore-wind-installed-country-on-track
-to-hit-wind-and-solar-targets-five-years-early-report-says.

40 **ultimately, the entire endeavor came to a halt:** Jonathan Shapiro,
"Federal Regulators Reject Appeal to Halt Nuclear Expansion at
Plant Vogtle," *Wabe*, April 16, 2012, wabe.org/federal-regulators
-reject-appeal-halt-nuclear-expansion-plant-vogtle.

42 **Instead, any geothermal energy:** "Regulatory Reform Could
Unlock Gigawatts of Zero Emissions Geothermal," ClearPath,
clearpath.org/tech-101/regulatory-reform-could-unlock-gigawatts
-of-zero-emission-geothermal.

43 **spotted owl habitats:** Jason Daley, "The Burning Questions About
Spotted Owls and Fire," *Sierra*, August 15, 2016, sierraclub.org
/sierra/2016-4-july-august/green-life/burning-questions-about
-spotted-owls-and-fire.

44 **"fundamental, substantive flaws":** Environmental and Energy
Law Program, eelp.law.harvard.edu/2017/09/defining-waters-of
-the-united-states-clean-water-rule

44 **In Georgia, the rule:** "Clean Water Rollbacks Enable Destruction
of 400 Acres of Wetlands," Southern Environmental Law Center,
October 26, 2020, southernenvironment.org/news/clean-water
-rollbacks-enable-destruction-of-400-acres-of-wetlands-near
-okefenokee.

45 **Climate policies such as:** "Explaining the Plummeting Cost of
Solar Power," MIT, November 20, 2018, news.mit.edu/2018
/explaining-dropping-solar-cost-1120.

46 **Car buyers must:** Shannon Osaka, "The EV Tax Credit Can Save
You Thousands—If You're Rich Enough," *Grist*, February 26, 2021,
grist.org/energy/the-ev-tax-credit-can-save-you-thousands-if-youre
-rich-enough.

47 **Because EVs are mostly:** Rachel Reed, "Current Electric Vehicles
Subsidies Fail to Reduce Overall Emissions, Says Harvard Law
Study," *Harvard Law Today*, April 7, 2022, hls.harvard.edu/today
/current-electric-vehicles-subsidies-fail-to-reduce-overall-emissions
-says-harvard-law-study.

47 **California has accepted:** "The Climate Commitment," RPrime
Foundation, 2023, climatecommitment.online/copy-of-moo.

47 **Meanwhile, all of California's:** Hayley Smith, "A Single,
   Devastating California Fire Season Wiped Out Years of Efforts to
   Cut Emissions," *Los Angeles Times*, October 20, 2022, latimes.com
   /california/story/2022-10-20/california-wildfires-offset-greenhouse
   -gas-reductions.

47 **Gavin Newsom is planning:** Aleks Phillips, "Gavin Newsom's
   California Electric Car Push Faces Huge Hurdles," *Newsweek*,
   January 7, 2023, newsweek.com/gavin-newsom-california-electric
   -vehicle-plan-climate-change-1771961.

47 **Norway is the only:** Kari Lundgren and Stephen Treloar, "Oil is
   Hard to Quit, Even in Norway Where Electric Cars Rule the
   Road," *Bloomberg*, July 6, 2023, bloomberg.com/news/articles
   /2023-07-07/oil-is-hard-to-quit-even-in-norway-where-electric-cars
   -rule-the-road?embedded-checkout=true.

54 **"Conservation will ultimately boil down to":** Terry Anderson,
   "Private Conservation in the Public Interest," *Perc*, December 2, 2015,
   perc.org/2015/12/02/private-conservation-in-the-public-interest.

Chapter Three | Unlocking America's Clean Future

56 **"subject to certain punishment":** "Beijing 2022 Official Warns
   Against Violations of 'Olympic Spirit,'" Reuters, January 19, 2022,
   reuters.com/lifestyle/sports/beijing-2022-official-says-athlete
   -protests-will-lead-punishment-2022-01-19.

57 **Several US climate-related:** "Meet the Deep-Pocketed Climate
   Nonprofit Pushing Gas Stove Ban with Direct Line to Biden
   Admin, China Links," Fox News, June 9, 2023, foxnews.com
   /politics/meet-deep-pocketed-climate-nonprofit-pushing-gas-stove
   -ban-direct-line-biden-admin-china-links.

58 **Venezuela's oil has been:** Maria Luisa Paul, "Venezuela's Oil
   Pollution: A Never-Ending Environmental Catastrophe," *The
   Washington Post*, October 7, 2021, washingtonpost.com/world
   /2021/10/07/oil-pollution-lake-maracaibo-venezuela.

58 **Russia's leaky, antiquated:** Paul Bledsoe, "Russian Gas Is a
   Climate and Security Disaster," *The Hill*, December 9, 2021,
   thehill.com/opinion/energy-environment/584956-russian-gas-is-a
   -climate-and-security-disaster.

60 **Very few people:** "Methane Matters," NASA Earth Observatory, March 8, 2016, earthobservatory.nasa.gov/features/MethaneMatters.

61 **To give you an idea:** Kevin Crowley and Ryan Collins, "Oil Producers Are Burning Enough 'Waste' Gas to Power Every Home in Texas," *Bloomberg*, April 11, 2019, bloomberg.com/news/articles /2019-04-10/permian-basin-is-flaring-more-gas-than-texas-residents -use-daily.

62 **Since 2007, US total emissions:** "Potent Greenhouse Gas Declines in the US," NOAA Research, March 3, 2023, research.noaa.gov/2023 /03/07/potent-greenhouse-gas-declines-in-the-us-confirming-success -of-control-efforts/#:~:text=Hu%20and%20her%20colleagues %20documented,and%20the%20EPA%20reporting%20requirement.

63 **But in fact, consumption-based:** "United States: CO2 Profile," Our World in Data, 2022, ourworldindata.org/co2/country /united-states.

64 **China may lead:** "Chinese Coal-Based Power Plants in the Belt and Road Initiative," Wilson Center, June 28, 2022, wilsoncenter.org /blog-post/chinese-coal-based-power-plants.

64 **In 2020, the Xi government:** "China Wants More Coal Power and to Hit Climate Change Targets," *Bloomberg*, October 31, 2022, bloomberg.com/news/articles/2022-10-31/china-wants-more-coal -power-and-to-hit-climate-change-targets.

64 **In fact, China is building:** "China's Coal Power Boom Cools Amid Climate Push," *The Wall Street Journal*, December 2, 2021, wsj.com/articles/chinas-coal-power-boom-beijing-xi-jinping -climate-energy-biden-administration-11650480857.

64 **In 2021, the country:** "China Wants More Coal Power and to Hit Climate Change Targets," *Bloomberg*, October 31, 2022, bloomberg.com/news/articles/2022-10-31/china-wants-more-coal -power-and-to-hit-climate-change-targets.

65 **In 2021, India:** "COP26: India PM Narendra Modi Pledges Net Zero by 2070," BBC, November 2, 2021, bbc.com/news/world-asia -india-59125143.

66 **In 2021, worldwide employment:** Stefan Ellerbeck, "The Renewable Energy Transition Is Creating a Green Jobs Boom," *GreenBiz*, January 20, 2023, greenbiz.com/article/renewable

-energy-transition-creating-green-jobs-boom#:~:text=12.7
%20million%20people%20work%20in,the%20International
%20Renewable%20Energy%20Agency.

67 **Global solar power–generating capacity:** Emiliano Bellini, "Global Solar Installations May Hit 350.6 GW in 2023, Says TrendForce," *PV Magazine*, February 16, 2023, pv-magazine.com /2023/02/16/global-solar-installations-may-hit-350-6-gw-in-2023 -says-trendforce.

67 **China currently controls:** "China's Solar Dominance Is Frustrating the U.S.," *The Washington Post*, July 30, 2022, washingtonpost.com /business/2022/07/30/climate-solar-manchin-china.

68 **Recent reports have indicated:** Aaron Mok, "Forced Uyghur Labor Is Being Used in China's Solar Panel Supply Chain, Researchers Say," *Business Insider*, November 30, 2022, businessinsider.com/forced-uyghur-labor-china-solar-panel-supply -chain-research-report-2022-11.

68 **In June 2022, US Customs:** "Uyghur Forced Labor Prevention Act," US Customs and Border Protection, July 21, 2023, cbp.gov /trade/forced-labor/UFLPA.

70 **Depending on available incentives:** "Reviewing the U.S. Solar Panel Value Chain Manufacturing Capacity," *PV Magazine*, March 7, 2023, pv-magazine-usa.com/2023/03/07/reviewing-the-u-s-solar -panel-value-chain-manufacturing-capacity.

70 **South Korea's Hanwha Solutions:** Joyce Lee and Heekyong Yang, "S. Korea's Hanwha Qcells to Invest $2.5 Bln in U.S. Solar Supply Chain," Reuters, January 11, 2023, reuters.com/business/energy /koreas-hanwha-qcells-invest-25-bln-us-solar-supply-chain-2023-01-11.

70 **Seven of the world's:** "China's Wind Turbine Manufacturing Holds 7 Spots Among World's Top 10," EVwind, March 13, 2021, evwind.es/2021/03/13/china-takes-up-7-spots-among-the-worlds -top-10-wind-turbine-manufacturers-for-wind-power/79787.

72 **Riverside East Solar Energy Zone:** Oliver Wainwright, "How Solar Farms Took Over the California Desert: An Oasis Has Become a Dead Sea," *The Guardian*, May 21, 2023, theguardian .com/us-news/2023/may/21/solar-farms-energy-power-california -mojave-desert.

73 **Interestingly, California recently:** Julie Cart, "California's Residential Solar Rules Overhauled After Highly Charged Debate," CalMatters, December 15, 2022, calmatters.org/environment /2022/12/california-solar-rules-overhauled.

75 **Every ton of lithium:** "How Much CO2 Is Emitted by Manufacturing Batteries?" MIT Climate Portal, July 15, 2022, climate.mit.edu/ask-mit/how-much-co2-emitted-manufacturing -batteries#:~:text=Particularly%20in%20hard%20rock% 20mining,are%20emitted%20into%20the%20air.

76 **Even under best-case scenarios:** "How Long Until Lithium Supply is Depleted?" DW, July 22, 2021, dw.com/en/is-e-mobility -going-to-crash-over-lithium-shortages/a-58214328.

76 **In a recent report, Bjørn:** Bjørn Lomborg, "Policies Pushing Electric Vehicles Show Why Few People Want One," *The Wall Street Journal*, September 9, 2022, wsj.com/articles/policies -pushing-electric-vehicles-show-why-few-people-want-one-cars -clean-energy-gasoline-emissions-co2-carbon-electricity -11662746452.

76 **Most cobalt, for instance:** "Congo Cobalt Mining for Lithium- Ion Battery," *The Washington Post*, September 30, 2016, washingtonpost.com/graphics/business/batteries/congo-cobalt -mining-for-lithium-ion-battery.

76 **The material prices:** Tim Levin, "Rising Battery Prices Keeping Electric Cars Expensive," *Business Insider*, December 7, 2022, businessinsider.com/cheap-electric-cars-delayed-ev-battery-prices -lithium-2022-12.

77 **By 2040, pure battery electric:** "Electric Vehicles Will Make Up the Majority of Car Sales by 2040," Energy5, August 28, 2023, energy5.com/electric-vehicles-will-make-up-the-majority-of-car -sales-by-2040.

78 **Experts estimate that China's:** Garrett Hering and J. Holzman, "US Lithium-Ion Battery Imports Jump as China Seizes Market Share," S&P Global, March 29, 2021, spglobal.com /marketintelligence/en/news-insights/latest-news-headlines/us -lithium-ion-battery-imports-jump-as-china-seizes-market-share -63271388.

79 **China has also taken:** Amit Katwala, "The World Can't Wean Itself Off Chinese Lithium," *Wired*, June 30, 2022, wired.com /story/china-lithium-mining-production.

79 **To meet the demand:** "Unlocking Critical Domestic Sources of Lithium for the UK," Northern Lithium, northernlithium. co.uk/#:~:text=Specifically%2C%20forecasts%20suggest% 20that%20the,electric%20vehicles%20and%20energy%20storage.

80 **To help solve this challenge:** Fred Lambert, "13 Battery Gigafactories Are Coming to the US by 2025," *Electrek*, December 27, 2021, electrek.co/2021/12/27/13-battery-gigafactories-coming -us-2025-ushering-new-era.

82 **The International Energy Agency admitted:** "Mineral Requirements for Clean Energy Transitions," International Energy Agency, 2022, iea.org/reports/the-role-of-critical-minerals-in-clean -energy-transitions/mineral-requirements-for-clean-energy -transitions.

86 **It produces about 35 percent:** "Russian and Chinese Designs in 87% of New Nuclear Reactors: IEA Chief," CNBC, July 1, 2022, cnbc.com/2022/07/01/russian-and-chinese-designs-in-87percent -of-new-nuclear-reactors-iea-chief.html.

86 **The Chinese government has approved:** Shunsuke Tabeta, "China Greenlights 6 New Nuclear Reactors in Shift Away from Coal," *Nikkei Asia*, October 29, 2021, asia.nikkei.com/Business/Energy /China-greenlights-6-new-nuclear-reactors-in-shift-away-from-coal.

87 **In the IEA's plan:** "Nuclear Power Can Play a Major Role in Enabling Secure Transitions to Low Emissions Energy Systems," IEA, June 30, 2022, iea.org/news/nuclear-power-can-play-a-major-role-in -enabling-secure-transitions-to-low-emissions-energy-systems.

87 **time of this writing, the US ranks:** Daniel Shats and Peter W. Singer, "The Balance of Power Is Shifting Among Nuclear Energy Titans," *Defense One*, October 31, 2022, defenseone.com/ideas /2022/10/balance-power-shifting-among-nuclear-energy-titans /378067.

88 **The amount of available geothermal energy:** "How Geothermal Energy Works," Union of Concerned Scientists, December 22, 2014, ucsusa.org/resources/how-geothermal-energy-works

97 **Texas leads the nation:** "Texas Profile," Energy Information Administration, June 15, 2023, eia.gov/state/?sid=TX#:~:text=Quick%20Facts,refining%20capacity%20in%20the%20nation.

98 **On top of this, Germany:** Philip Oltermann, "How Reliant is Germany on Russian Gas?" *The Guardian*, July 21, 2022, theguardian.com/world/2022/jul/21/how-reliant-is-germany-and-europe-russian-gas-nord-stream.

99 **45 percent of all American coal:** Will Wade, "U.S. Coal Power to Fall 45% by Decade's End in Shift from Dirtiest Fossil Fuel," *Bloomberg*, April 4, 2022, bloomberg.com/news/articles/2022-04-04/u-s-coal-power-to-fall-45-by-decade-s-end-in-fossil-fuel-shift.

102 **In a 2022 report sponsored:** "DOE Report Finds Hundreds of Retiring Coal Plant Sites Could Convert to Nuclear," US Department of Energy, September 13, 2022, energy.gov/ne/articles/doe-report-finds-hundreds-retiring-coal-plant-sites-could-convert-nuclear.

102 **A single SMR:** "New Analysis Shows Economic Benefits in a Coal to Nuclear Transition," Nuclear Energy Institute, October 28, 2021, nei.org/news/2021/analysis-shows-economic-benefit-in-coal-to-nuclear.

104 **The actor touts:** Jill Serjeant, "DiCaprio Calls 'Don't Look Up' a 'Unique Gift' to Climate Change Fight," Reuters, December 8, 2021, reuters.com/lifestyle/dicaprio-calls-dont-look-up-unique-gift-climate-change-fight-2021-12-08.

104 **Add to it blatant hypocrisy:** Rupert Neate, "Super-Rich Fuelling Growing Demand for Private Jets, Report Finds," *The Guardian*, October 27, 2019, theguardian.com/environment/2019/oct/27/super-rich-fuelling-growing-demand-for-private-jets-report-finds.

106 **They also added this message:** Colin Moynihan, "The Climate Clock Now Ticks with a Tinge of Optimism," *The New York Times*, April 19, 2021, nytimes.com/2021/04/19/arts/design/climate-change-clock-new-york.html.

106 **For example, the United States:** "Ranking 41 US States Decoupling Emissions and GDP Growth," World Resources

Institute, July 28, 2020, wri.org/insights/ranking-41-us-states
-decoupling-emissions-and-gdp-growth.

107 **South Dakota now produces:** "South Dakota Is Building a Clean
Energy Future," Clean Grid Alliance, November 21, 2022,
cleangridalliance.org/blog/136/south-dakota-is-building-a-clean
-energy-future#:~:text=South%20Dakota%20now%20produces
%20twice,over%20the%20next%20five%20years.

110 **In the United States today:** "Facts and Figures About Materials,
Waste, and Recycling," US Environmental Protection Agency, epa
.gov/facts-and-figures-about-materials-waste-and-recycling
/national-overview-facts-and-figures-materials.

110 **In 1800, more than:** "Farm Labor," US Department of Agriculture,
August 7, 2023, ers.usda.gov/topics/farm-economy/farm-labor.

111 **The EPA estimates that agriculture:** "Sources of Greenhouse Gas
Emissions," US Environmental Protection Agency, August 25,
2023, epa.gov/ghgemissions/sources-greenhouse-gas-emissions.

111 **Because of such innovations:** "Agriculture," US Environmental
Protection Agency, 2017, epa.gov/sites/default/files/2017-02
/documents/2017_chapter_5_agriculture.pdf.

112 **Taking into account:** "Investing in Public R&D for a Competitive
and Sustainable US Agriculture," Breakthrough Institute, March
16, 2021, thebreakthrough.org/issues/food-agriculture
-environment/investing-in-r-d-for-us-ag.

117 **By working with local:** "15 Million Acres and Counting," Ducks
Unlimited, May 12, 2021, ducks.org/conservation/national
/15-million-acres-and-counting.

118 **ConocoPhillips owns 643,000:** "Wetlands Coastal,"
ConocoPhillips, 2018, static.conocophillips.com/files/resources
/17-0488-coastal-wetlands-brochure-022018-comp1.pdf.

119 **These phytoplankton absorb:** S. Irion, U. Christaki, H. Berthelot,
et al., "Small Phytoplankton Contribute Greatly to CO2-Fixation
after the Diatom Bloom in the Southern Ocean," ISME J 15,
2509–2522 (2021), doi.org/10.1038/s41396-021-00915-z.

119 **With approximately 2,500 miles:** "Can We Use CO2 in a
Beneficial Way?" Occidental Petroleum Corporation, oxy.com
/operations/performance-production/eor.

123 **For example, since 2005:** Robert Rapier, "Yes, the U.S. Leads All
Countries in Reducing Carbon Emissions," *Forbes*, October 24,
2017, forbes.com/sites/rrapier/2017/10/24/yes-the-u-s-leads-all
-countries-in-reducing-carbon-emissions/?sh=17ef03343535.

126 **In 2021, for example:** "More Than One Quarter of All Venture
Capital Funding is Going to Climate Technology, with Increased
Focus on Technologies That Have the Most Potential to Cut
Emissions," PwC, March 11, 2023, pwc.com/gx/en/news-room
/press-releases/2022/state-of-climate-tech-report-2022.html.

127 **While not perfect:** "Inflation Reduction Act Guidebook," The
White House, 2023, whitehouse.gov/cleanenergy/inflation
-reduction-act-guidebook.

130 **In 2021 alone:** "State of the Outdoor Market," Outdoor Industry
Association, December 2022, outdoorindustry.org/wp-content
/uploads/2022/12/OIA-State-of-the-Outdoor-Market-Report-Fall
-2022.pdf.

134 **Since then, 186:** "ESG on the Edge: Controversy Weighs on
Sustainable ETFs," CNBC, September 11, 2022, cnbc.com/2022
/09/11/esg-on-the-edge-controversy-weighs-on-sustainable-etfs.html.

135 **In October 2022, nineteen:** "Six Big Banks Investigated Over
Net-Zero Emissions Pledge," PBS, March 2, 2023, pbs.org/wnet
/peril-and-promise/2023/03/six-big-banks-investigated-over-net
-zero-emissions-pledge/#:~:text=In%20October%20of%202022
%2C%20nineteen,Zero%20Banking%20Alliance%20(NZBA).

135 **For instance, Anheuser-Busch's ESG:** "The ESG Empire Strikes
Back Following Bud Light Boycott," Foundation for Economic
Education (FEE), May 31, 2023, fee.org/articles/the-esg-empire
-strikes-back-following-bud-light-boycott.

136 **Natural gas extracted:** Mike Soraghan and Carlos Anchondo,
"Does a Crackdown on Russian Gas Help or Hurt the Climate?"
E&E News by *Politico*, May 13, 2022, eenews.net/articles/does-a
-crackdown-on-russian-gas-help-or-hurt-the-climate.

137 **In a 2020 Cambridge:** "New Survey Reveals 39% of U.S. Students
Believe Climate Change the Most Pressing Issue Facing the World

Today," Cambridge Assessment International Education, March 3, 2020, cambridgeinternational.org/news/news-details/view /new-survey-reveals-39-percent-of-us-students-believe-climate -change-the-most-pressing-issue-facing-the-world-today-2020 0303.

139 **Important projects at Texas A&M:** "Table-to-Farm Approach Will Create Healthier U.S. Melon Supply chain," AgriLife, April 13, 2021, agrilifetoday.tamu.edu/2021/04/13/table-to-farm -approach-will-create-healthier-u-s-melon-supply-chain.

141 **the US nuclear industry already supports:** Christopher Barnard, "The Global Nuclear Comeback: Green Energy, Fossil Fuels Supply Climate Mandates Power Generation," *The Wall Street Journal*, July 18, 2022, wsj.com/articles/the-global-nuclear-comeback-green -energy-fossil-fuels-supply-climate-mandates-power-generation -11658170860?mod=opinion_lead_pos6.

142 **The US nuclear industry also:** "Towards a Just Energy Transition: Nuclear Power Boasts Best Paid Jobs in Clean Energy Sector," International Atomic Energy Agency (IAEA), April 14, 2022, iaea .org/newscenter/news/towards-a-just-energy-transition-nuclear -power-boasts-best-paid-jobs-in-clean-energy-sector.

Chapter Six | Conserving Our Environment, Our Heritage, and Our Future

147 **It is true that a loss:** "Nature's Make or Break Potential For Climate Change," The Nature Conservancy, October 15, 2017, nature.org/en-us/what-we-do/our-insights/perspectives/natures -make-or-break-potential-for-climate-change.

148 **In 2020 alone, California forest:** Elena Shao, "Mapping California's 'Zombie' Forests," *The New York Times*, March 6, 2023, nytimes.com/interactive/2023/03/06/climate/california -zombie-forests.html.

148 **The World Economic Forum also reported:** "This is How Much Carbon Wildfires Have Emitted This Year," World Economic Forum (WEF), December 10, 2021, weforum.org/agenda/2021/12 /siberia-america-wildfires-emissions-records-2021.

149 **With 85 percent:** "Wildfire Causes and Evaluations," U.S. National Park Service, 2018, nps.gov/articles/wildfire-causes-and -evaluation.htm.

150 **More than 150,000 Canadians:** Sue Allan and Nick Taylor-Vaisey, "'Literally off the Charts': Canada's Fire Season Sets Records—and Is Far from Over," *Politico*, July 6, 2023, politico.com/news/2023 /07/06/canada-fire-season-00104959.

150 **Cities such as New York:** Emma G. Fitzsimmons and Luis Ferré-Sadurní, "Did N.Y. Leaders Leave Residents Unprepared for the Air Quality Crisis?" June 9, 2023, *The New York Times*, nytimes .com/2023/06/09/nyregion/adams-hochul-response-air-quality.

151 **Current data shows that conservation:** "Natural Climate Solutions," American Conservation Coalition, 2021, acc.eco/s /Natural-Climate-Solutions-Booklet-Final-actually.pdf.

152 **Take Alaska, where:** Carly A. Phillips et al., "Escalating Carbon Emissions from North American Boreal Forest Wildfires and the Climate Mitigation Potential of Fire Management," *Sci. Adv.* 8, no. 17, April 27, 2022, doi.10.1126/sciadv.abl7161.

152 **If, however, that state:** "The Climate Commitment," RPrime Foundation, 2023, climatecommitment.online.

153 **In 2018 alone, global tree:** Carly A. Phillips et al., "Natural Climate Solutions," American Conservation Coalition, 2021, acc .eco/s/Natural-Climate-Solutions-Booklet-Final-actually.pdf.

156 **One recent study shows:** Carly A. Phillips, et al., "Escalating Carbon Emissions from North American Boreal forest wildfires and the climate mitigation potential of fire management," *Sci. Adv.* 8, no. 17, April 2022, doi.10.1126/sciadv.abl7161.

156 **Back in 1911:** "Fire Behavior on the Stanislaus National Forest," United States Forest Service, 2014, fs.usda.gov/detail/stanislaus /home/?cid=stelprd3808225.

158 **Once upon a time:** "Hydrologic Restoration of Coastal Wetlands Enhances Climate Change Mitigation," US Geological Survey (USGS), September 29, 2022, usgs.gov/centers/whcmsc/news /hydrologic-restoration-coastal-wetlands-enhances-climate-change -mitigation.

162 **According to Restore America's Estuaries:** "Coastal Blue Carbon," Restore America's Estuaries, 2018, estuaries.org/wp-content/uploads /2018/08/RAE-Blue-Carbon-FactSheet_9March2018.pdf.

164 **To date, about two thousand:** "Amazon Deforestation and Climate Change," *National Geographic*, September 19, 2023, national geographic.org/media/amazon-deforestation-and-climate-change.

165 **Each year, forty-seven million:** "Hunters and Anglers," Theodore Roosevelt Conservation Partnership, trcp.org/images/uploads /wygwam/Econ-fact-sheet.pdf.

166 **Since its founding in 1937:** "America's River Initiative," Ducks Unlimited, ducks.org/conservation/du-conservation-initiatives /americas-river-initiative.

168 **Sadly, well over 50 percent:** Jennifer Gray, "Report: Marine Life Has Taken a Devastating Hit Over 40 Years," CNN, September 17, 2015, cnn.com/2015/09/17/world/oceans-report/index.html

169 **Coral reefs, although they now:** United Nations, un.org/en /observances/oceans-day/360diving.

171 **In the last five years:** "Sustainability," American Farm Bureau Federation, fb.org/issue/sustainability.

173 **Besides preventing the release:** "No-Till Farming Improves Soil Health and Mitigates Climate Change," Environmental and Energy Study Institute (EESI), March 28, 2022, eesi.org/articles/view /no-till-farming-improves-soil-health-and-mitigates-climate-change.

174 **Macauley Farms, which produces beef:** "Soil Health Case Study," American Farmland Trust, December 2020, farmlandinfo.org /wp-content/uploads/sites/2/2020/12/NY_MacauleyFarms_Soil _Health_Case_Study_AFT_NRCS.pdf.

174 **That's about 1.3 billion:** Sarah Kaplan, "Climate Curious: Food Waste," *The Washington Post*, February 25, 2021, washingtonpost .com/climate-solutions/2021/02/25/climate-curious-food-waste.

174 **In the US alone, wasted:** "Fight Climate Change by Preventing Food Waste," World Wildlife Fund, worldwildlife.org/stories /fight-climate-change-by-preventing-food-waste.

175 **Studies estimate that if we stopped:** Clare McCarthy, "Don't Table Food Waste in Climate Conversations!" Emory University,

November 23, 2021, climatetalks.emory.edu/blog/dont-table-food
-waste-climate-conversations.

175 **The 28 percent:** Sarah Kaplan, "Climate Curious: Food Waste,"
The Washington Post, February 25, 2021, washingtonpost.com
/climate-solutions/2021/02/25/climate-curious-food-waste.

175 **A whopping 8 percent of global:** "Food Waste Is Contributing to
Climate Change, What's Being Done About It?" PBS, November
26, 2022, pbs.org/newshour/show/food-waste-is-contributing-to
-climate-change-whats-being-done-about-it.

Chapter Seven | Innovating Our Way Out of the Energy Crisis

181 **The Hydrogen Council estimates:** "Hydrogen—Scaling Up,"
The Hydrogen Council, 2017, hydrogencouncil.com/wp-content
/uploads/2017/11/Hydrogen-Scaling-up_Hydrogen-Council_2017
.compressed.pdf.

183 **Texas is a state in which:** "Texas State Energy Profile," Energy
Information Administration, June 15, 2023, eia.gov/state/print
.php?sid=TX.

184 **Besides cost, the reliability:** "2021 State of the Market Report for
the ERCOT Electricity Markets," Potomac Economics, May 2022,
p. 17, potomaceconomics.com/wp-content/uploads/2022/05
/2021-State-of-the-Market-Report.pdf.

186 **As many of us learned:** "Laws of Energy," Energy Information
Administration, eia.gov/kids/what-is-energy/laws-of-energy.

186 **And as global energy demand:** "Energy Production and
Consumption," Our World in Data, ourworldindata.org/energy
-production-consumption.

188 **The current number of people:** "Access to Electricity,"
International Energy Agency, 2023, iea.org/reports/sdg7-data-and
-projections/access-to-electricity.

188 **Therefore, it should come:** "Fossil Fuels, Explained," *National
Geographic*, April 2, 2019, nationalgeographic.com/environment
/article/fossil-fuels.

189 **latest hardware and software:** "GE Power to Turn Old Power
Plant into Technology Center," Power-Grid International, May 6,

2017, power-grid.com/energy-efficiency/ge-power-to-turn-old
-power-plant-into-technology-center/#gref.

190 **The Ljungström factory:** Jimmy Vielkind, "A New York Town
Once Thrived on Fossil Fuels, Now Wind Energy Is Giving a Lift,"
The Wall Street Journal, March 5, 2023, wsj.com/articles/a-new
-york-town-once-thrived-on-fossil-fuels-now-wind-energy-is-giving
-a-lift-d693433a.

190 **Because nuclear energy already:** "Nuclear Power in the USA,"
World Nuclear Association, August 2023, world-nuclear.org
/information-library/country-profiles/countries-t-z/usa-nuclear
-power.aspx.

190 **Finally, employment opportunities:** "Can Advanced Nuclear
Repower Coal Country?" Bipartisan Policy Center, March 23,
2023, bipartisanpolicy.org/report/nuclear-repower-in-coal-country.

193 **Right now, annual use:** Jade Boutot, Adam S. Peltz, Renee McVay,
Mary Kang, "About CCUS," International Energy Agency, April
2021, iea.org/reports/about-ccus.

193 **Some studies estimate that:** "Documented Orphaned Oil and Gas
Wells across the United States (2021 & 2022 Coordinates)," 2022,
doi.org/10.5683/SP3/PLAOIX.

195 **Even after the current federal:** "How Much Do Solar Panels Cost
in 2023?" Energy Sage, May 1, 2023, energysage.com/local-data
/solar-panel-cost.

195 **In California, residents pay:** "Why Is Electricity So Expensive in
California?" Ivy Energy, September 7, 2022, ivy-energy.com/blog
/californias-ever-increasing-electricity-costs.

196 **In addition, these panels:** Brian Gitt, "Solar's Dirty Secrets: How
Solar Power Hurts People and the Planet," October 8, 2021, briangitt
.com/solars-dirty-secrets-how-solar-power-hurts-people-and-the-planet.

196 **One recent study done:** Sammy Roth, "Solar and Wind Farms
Can Hurt the Environment: A New Study Offers Solutions," *Los
Angeles Times*, October 6, 2022, latimes.com/environment
/newsletter/2022-10-06/solar-and-wind-farms-can-hurt-the
-environment-a-new-study-offers-solutions-boiling-point.

197 **Similarly, to paint a clearer picture:** RPrime Foundation, 2023,
climatecommitment.online.